SpringerBriefs in Electrical and Computer Engineering

Computational Electromagnetics

Series editors

K.J. Vinoy, Bangalore, India
Late Rakesh Mohan Jha, Bangalore, India

More information about this series at http://www.springer.com/series/13885

Hema Singh · Simy Antony
Harish Singh Rawat

EM Wave Propagation Analysis in Plasma Covered Radar Absorbing Material

Hema Singh
Centre for Electromagnetics (CEM)
CSIR-National Aerospace Laboratories
Bengaluru, Karnataka
India

Simy Antony
Centre for Electromagnetics (CEM)
CSIR-National Aerospace Laboratories
Bengaluru, Karnataka
India

Harish Singh Rawat
Apex Level Standards & Industrial
  Metrology
CSIR-National Physical Laboratory
New Delhi
India

ISSN 2191-8112   ISSN 2191-8120   (electronic)
SpringerBriefs in Electrical and Computer Engineering
ISSN 2365-6239   ISSN 2365-6247   (electronic)
SpringerBriefs in Computational Electromagnetics
ISBN 978-981-10-2268-5   ISBN 978-981-10-2269-2   (eBook)
DOI 10.1007/978-981-10-2269-2

Library of Congress Control Number: 2016947381

Printed on acid-free paper

This Springer imprint is published by Springer Nature
The registered company is Springer Science+Business Media Singapore Pte Ltd.

To Late Dr. R.M. Jha

Preface

Radar cross section (RCS) reduction can be achieved by controlling the reflections from the surface of the structure. Plasma envelope is one of the ways to control the reflections and scattering from the surface. The problem of electromagnetic (EM) propagation within bounded plasma can be approximated as multilayered dielectric problem. In this book, EM wave propagation within the plasma covered radar absorbing material (RAM) is discussed. The analytical formulation for the reflection coefficient of the plasma covered RAM based on impedance transformation method is presented. Both homogeneous and inhomogeneous plasma are considered. The effect of plasma parameters, such as electron density, collision frequency, plasma thickness, plasma density profile, etc., on the absorption behavior of plasma–RAM structure is discussed through various illustrations. This book provides an insight of EM propagation within plasma, that is, the basis of achieving plasma-based stealth.

Bengaluru, India

Hema Singh
Simy Antony
Harish Singh Rawat

Acknowledgments

We would like to thank Mr. Jitendra J. Jadhav, Director, CSIR-National Aerospace Laboratories, Bangalore for the permission to write this SpringerBriefs.

We would also like to acknowledge valuable suggestions from our colleagues and project staff at the Centre for Electromagnetics and their invaluable support during the course of writing this book.

But for the concerted support and encouragement of our Springer editorial contacts, it would not have been possible to bring out this book within such a short span of time. We very much appreciate the continued support extended by Springer.

Contents

About the Authors

Dr. Hema Singh is currently working as Principal Scientist in *Centre for Electromagnetics* of CSIR-National Aerospace Laboratories, Bangalore, India. Earlier, she was Lecturer in EEE, BITS, Pilani, India during 2001–2004. She obtained her Ph.D. degree in Electronics Engineering from IIT-BHU, Varanasi India in 2000. Her active area of research is Computational Electromagnetics for Aerospace Applications. More specifically, the topics she has contributed to, are GTD/UTD, EM analysis of propagation in an indoor environment, Phased Arrays, Conformal Antennas, Radar Cross Section (RCS) Studies including Active RCS Reduction. She received Best Woman Scientist Award in CSIR-NAL, Bangalore for period of 2007–2008 for her contribution in the area of phased antenna array, adaptive arrays, and active RCS reduction. Dr. Singh has coauthored nine books, one book chapter, and over 200 scientific research papers and technical reports.

Ms. Simy Antony obtained B.Tech. (ECE) from University of Calicut, India and M.Tech. in Electronics (Microwave and Radar Electronics) from Cochin University of Science and Technology, Kerala, India. She was a Project Scientist at the Centre for Electromagnetics (CEM) of CSIR-National Aerospace Laboratories, Bangalore where she worked on RCS studies for aerospace vehicles.

Harish Singh Rawat obtained M.Sc. (Electronics) in 2013 from Department of Engineering and Technology, Jamia Millia Islamia, New Delhi, India. He worked as Project Engineer at Centre for Electromagnetics (CEM) of CSIR-National Aerospace Laboratories, Bangalore, India. Currently, he is doing Ph.D. in CSIR-NPL, New Delhi.

His research interests include RCS studies, radar absorbing structures (RAS), low-profile antennas, phased arrays mounted on planar and nonplanar surfaces.

About the Book

This book focuses on EM propagation characteristics within multilayered plasma-dielectric-metallic media. The method used for analysis is impedance transformation method. Plasma covered radar absorbing material is approximated as a multilayered dielectric medium. The plasma is considered to be bounded homogeneous/inhomogeneous medium. The reflection coefficient and hence return loss is analytically derived. The role of plasma parameters, such as electron density, collision frequency, plasma thickness, and plasma density profile in the absorption behavior of multilayered plasma–RAM structure is described. This book provides a clearer picture of EM propagation within plasma. The reader will get an insight of plasma parameters that play significant role in deciding the absorption characteristics of plasma covered surfaces.

List of Figures

List of Tables

EM Wave Propagation Analysis in Plasma-Covered Radar Absorbing Material

Abstract The radar cross section (RCS) reduction can be achieved by controlling the reflections from the surface of the structure. Plasma envelope is one of the ways to control the reflections and scattering from the surface. The problem of electromagnetic (EM) propagation within bounded plasma can be approximated as multilayered dielectric problem. In this book, EM wave propagation within the plasma-covered radar absorbing material (RAM) is discussed. The analytical formulation for the reflection coefficient of the plasma-covered RAM based on impedance transformation method is presented. Both homogeneous and inhomogeneous plasma are considered. The effect of plasma parameters, such as electron density, collision frequency, plasma thickness, plasma density profile, etc. on the absorption behavior of plasma-RAM structure is discussed through various illustrations. This book provides an insight of EM propagation within plasma that is the basis of achieving plasma-based stealth.

Keywords EM wave propagation · Plasma · Radar absorbing structure · Radar cross section · Absorption

1 Introduction

Radar cross section (RCS) of a target is the measure of its detectability. It is related to the scattering characteristics of the target, depending on the incident wave frequency, target shape, material, and orientation of the target *w.r.t.* the incident wave. The target having metallic components are easy to be detected by the RADAR due to the high reflections. Thus, it becomes important to design the targets based on low-observable principle for reducing the reflections from the conductive surface. This may be achieved using *radar absorbing structures* (RAS), *radar absorbing materials* (RAM), or plasma apart from the shaping methods.

In this book, the EM wave propagation within plasma covered RAM is studied. The objective of the study is toward control of reflections from the surface. The reflection coefficient of plasma-covered RAM is derived. Both homogeneous and

© The Author(s) 2017
H. Singh et al., *EM Wave Propagation Analysis in Plasma Covered Radar Absorbing Material*, SpringerBriefs in Computational Electromagnetics, DOI 10.1007/978-981-10-2269-2_1

inhomogeneous plasmas are considered. The analysis of EM propagation within the plasma is complex due to its peculiar characteristics. The propagation behavior within plasma depends on the plasma parameters, such as electron density, collision frequency, plasma thickness, plasma density profile, etc. The problem of EM propagation within bounded plasma can be approximated as multilayered dielectric problem. For such multilayered structure, impedance transformation method (Hayt and Buck 2011) is one of the commonly used methods to analyze the reflection behavior of the structure.

The total reflection from the plasma covered conducting structure includes reflection at the plasma–air interface, partial reflection within the plasma layer, and the reflections from the conductor. The reflections from the conducting surface can be reduced by covering it with RAM. However, this RAM covering will not cancel out reflections entirely. The portion of incident wave that has transmitted through plasma will be incident and hence absorbed by RAM layer. There will be some reflections at the plasma–air interface. In case of open plasma, problems such as plasma trail, visible glow, etc., limits its application in stealth techniques. In order to avoid such problems, a plastic envelope is used to cover the plasma. An impedance transformation method is employed to study the reflections from bounded homogeneous/inhomogeneous plasmas covering RAM structure. Similar problem has been studied for homogeneous plasma by Yuan et al. (2011). It is noted that inhomogeneous plasma has better performance as compared to homogeneous plasma (Zhengli et al. 2010). This is because of the gradual impedance transition due to the inhomogeneous nature of the plasma (density profile) at the air–plasma interface. In this document, linear, parabolic, and exponential electron density profiles are considered for inhomogeneous plasma.

This book is divided into five sections. Section 2 describes the plasma parameters that can be used to control the EM wave reflection from the plasma layer. Section 3 describes the theoretical formulation for EM propagation in the bounded plasma. Equations are derived for only bounded plasma and RAM structure covered by bounded plasma with different density profiles. Section 4 discusses the computed results and analysis. Section 5 concludes the study carried out.

2 Role of Plasma Parameters

The low observability of the target can be achieved by reducing the reflections from its surface. Plasma is one of the choices for controlling the reflections from the surface. In order to use plasma for stealth-related applications, it should be designed as an absorber. Since plasma is essentially cloud of ionized particles, the reflection of incident EM wave depends up on its interactions with the plasma electrons. Depending up on the nature of plasma electrons, it can behave as an absorber or a reflector.

The important plasma parameters that affect the plasma performance are plasma frequency (ω_p), electron density (N_e), electron density profile, collision frequency

(V_{en}), and the plasma thickness (d). The variation in these parameters changes the dielectric constant of the plasma and hence its impedance.

The dielectric constant of the plasma can be expressed as (Fridman and Kennedy 2004)

$$\varepsilon = \left(1 - \frac{\omega_p^2}{\omega^2 + V_{en}^2}\right) - j \times \frac{V_{en}}{\omega} \frac{\omega_p^2}{\omega^2 + V_{en}^2} \tag{1}$$

where ω_p is the plasma frequency, V_{en} is the collision frequency, and ω is the incident wave frequency.

Electron density (N_e): The EM wave can enter into the plasma provided its wavelength is comparable with the average separation between the plasma electrons. If the electron density is high, the average distance between the plasma electrons will be small. Due to this decreased average distance between electrons, the incident EM wave is required to have smaller wavelength or higher frequency in order to enter into the plasma. The electron density plays an important role in determining the cut-off frequency in the plasma medium (Jenn 2005).

Plasma frequency (ω_p): Plasma frequency is determined by the electron density (N_e) of the plasma medium. It acts as a cut-off frequency for EM wave in lossless plasma medium. However, this is not valid for lossy plasma medium.

If the electron density of plasma is high, the nature of the plasma will be close to that of conductor. In other words, the plasma will act as a reflector.

The plasma frequency is expressed as (Bittencourt 2004)

$$\omega_p = \sqrt{\frac{N_e e^2}{m_e \varepsilon_0}} \tag{2}$$

where N_e is the plasma electron density, e is the electron charge, m_e is the electron mass, and ε_0 is the free space permittivity.

Collision frequency (V_{en}): Collisions within the plasma is responsible for the absorption mechanism in the plasma. The plasma is categorized as cold and hot plasmas, depending on plasma temperature. The cold and dense plasma is known to be strongly coupled plasma. The collisions are more effective in case of strongly coupled plasma. For weakly ionized plasma, electron collisions with neutral atoms will dominate over other collisions. The collision frequency in the plasma is determined by (Shul and Pearton 2000)

$$V_{en} = N_g \sigma_e V_{av} \tag{3}$$

where N_g is the density of gas molecules, σ_e is the cross section of electron elastic collisions, and V_{av} is the average velocity of electron.

The collision frequency affects the dielectric constant of the plasma, and makes the plasma a lossy medium. As EM wave propagates into the plasma, plasma electrons oscillate that are already in collisions with the neutral atoms. These collisions result in transfer of energy between electrons and the neutral atoms. This energy transfer will be maximum when the frequency of incident EM wave is close to the collision frequency of the plasma.

Plasma thickness (*d*): Bounded plasma gives rise to multiple reflections at plasma–air interface, reflections from the conductor, reflections in between these two surfaces. The bouncing of wave within the plasma medium can be considered as cavity resonance effect. At particular plasma thickness, the phases of the reflected waves in the plasma cancel each other, thereby reducing the effect of reflection. The condition for the cavity resonator is given by

$$2d_c = N\lambda; \quad N = 1, 2, 3, \ldots \tag{4}$$

where d_c is the thickness of the cavity and λ is the wavelength of incident EM wave in the medium. The cavity resonance effect in the plasma is different from that of ideal cavity resonator. This is mainly due to the finite resistance of the plasma boundaries unlike PEC boundary of cavity resonator.

Electron density profile: For inhomogeneous plasma, electron density is not uniform. The density gradient in the plasma varies the plasma frequency. Moreover due to electron density gradient, the dielectric constant of the plasma varies. If the electron density profile has gradual increase of density from the boundary toward the conductor, reflection at air–plasma interface will be less. Different electron density profiles such as linear (Mo and Yuan 2008), exponential, parabolic (Tang et al. 2003), Epstein (Chaudhury and Chaturvedi 2005) have been used for plasma-based studies.

3 Formulation for EM Propagation in Plasma-Covered RAM

In this book, bounded plasma has been considered for analyzing EM propagation within the plasma. The EM wave, propagating in—z direction, is assumed to normally incident on the plasma layer. The reflections of EM wave at the interfaces (air–plasma, plasma–conductor) are determined in terms of layer parameters, such as thickness, permittivity, and permeability of each layer. The impedance of each layer depends on these layer parameters. The impedance transformation method is used considering the conductor-backed plasma as multilayered dielectric structure. The impedance is transformed as EM wave propagates from air to plasma toward the conductor. Here first two-layered structure consisting of plasma and conductor is considered for the analysis of reflection phenomena. Next the homogeneous and inhomogeneous bounded plasma covering RAM structure is taken for propagation analysis.

3.1 Two-Layered Structure: Bounded Plasma

The permittivity and permeability of plasma layer depends on the incident wave frequency (ω), electron density (N_e), and collision frequency (V_{en}). The envelope that covers plasma is assumed as lossless and thin medium, thus its effect can be neglected. The parameters, Z_1, Z_2, k_1, k_2, μ_1, μ_2, ε_1, ε_2, are the impedance, wave number, relative permeability, and relative permittivity in air and plasma layer, respectively. Figure 1 shows the two-layered structure for the analysis.

In Layer 1 (air), the impedance is given by

$$Z_1 = Z_0 \tag{5a}$$

$$k_1 = \frac{2\pi f\left(\sqrt{\mu_1 \varepsilon_1}\right)}{c} \tag{5b}$$

where $\mu_1 \varepsilon_1 = \mu_0 \varepsilon_0$, Z_0 is the free space impedance and c is the speed of the light in free space.

The electric and magnetic fields in Region 1 are expressed as

$$E_y = E_1^+ \exp(jk_1 z) + E_1^- \exp(-jk_1 z) \tag{6a}$$

$$H_x = \frac{E_1^+ \exp(jk_1 z) - E_1^- \exp(-jk_1 z)}{Z_1} \tag{6b}$$

In Layer 2 (plasma), the impedance is given by

$$Z_2 = \sqrt{\frac{\mu_2}{\varepsilon_2}} Z_0 \tag{7a}$$

$$k_2 = \frac{2\pi f\left(\sqrt{\mu_2 \varepsilon_2}\right)}{c} \tag{7b}$$

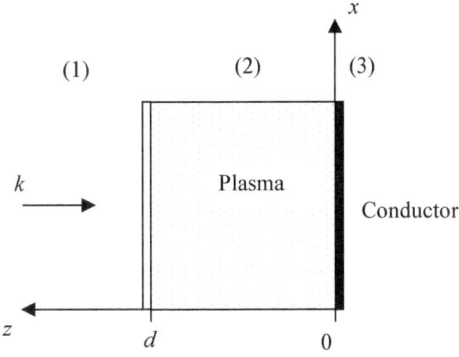

Fig. 1 Two-layered structure: conductor-backed bounded plasma

The parameters, μ_2, ε_2 depends on the nature of the medium. For nonmagnetic plasma, μ_2 and ε_2 are obtained as

$$\mu_2 = 1 \tag{8a}$$

$$\varepsilon_2 = \left(1 - \frac{\omega_p^2}{\omega^2 + V_{en}^2}\right) - j \times \frac{V_{en}}{\omega} \frac{\omega_p^2}{\omega^2 + V_{en}^2} \tag{8b}$$

The electric and magnetic fields in plasma are expressed as

$$E_y = E_2^+ \exp(jk_2 z) + E_2^- \exp(-jk_2 z) \tag{9a}$$

$$H_x = \frac{E_2^+ \exp(jk_2 z) - E_2^- \exp(-jk_2 z)}{Z_2} \tag{9b}$$

The Layer 3 is assumed as conductor. It will reflect the incident wave, i.e.,

$$\Gamma_{23} = -1$$

The wave impedance in the second layer is given by

$$\begin{aligned}
Z(z) &= \frac{E_x(2)}{H_y(2)} = \frac{E_2^+ \exp(jk_2 z) + \Gamma_{23} E_2^- \exp(-jk_2 z)}{E_2^+ \exp(jk_2 z) - \Gamma_{23} E_2^- \exp(-jk_2 z)} \times Z_2 \\
&= \frac{E_2^+ \exp(jk_2 z) - E_2^+ \exp(-jk_2 z)}{E_2^+ \exp(jk_2 z) + E_2^+ \exp(-jk_2 z)} \times Z_2 \\
&= \frac{\exp(jk_2 z) - \exp(-jk_2 z)}{\exp(jk_2 z) + \exp(-jk_2 z)} \times Z_2 \\
&= Z_2 \tanh jk_2 z
\end{aligned} \tag{10}$$

Applying boundary conditions at air–plasma interface, one gets

$$E_y(1) = E_2^+ ; \quad z = d \tag{11a}$$

$$H_x(1) = H_2^+ ; \quad z = d \tag{11b}$$

where, d is the thickness of the plasma layer.
    Equations (11a) and (11b) can be written as

$$E_1^+ + E_1^- = E_2^+ \tag{12a}$$

$$\frac{E_1^+ - E_1^-}{Z_1} = \frac{E_2^+}{Z(d)} \tag{12b}$$

where $\frac{E_2^+}{H_2^+} = Z(d)$

Dividing (12a) by (12b), one gets

$$Z_1 \frac{E_1^+ + E_1^-}{E_1^+ - E_1^-} = Z(d) \tag{13}$$

where $Z(d)$ is the impedance beyond the first interface.

Using (10), one has

$$\begin{aligned} Z(d) &= Z_2 \tanh jk_2 d \\ &= Z_2 \delta_2 \end{aligned} \tag{14}$$

where $\delta_2 = \tanh jk_2 d$.

Using (14), the Eq. (13) can be reduced to

$$\frac{E_1^-}{E_1^+} = \frac{Z_2 \delta_2 - Z_1}{Z_2 \delta_2 + Z_1} = \Gamma \tag{15}$$

where Γ is the reflection coefficient of the two-layered conductor-backed plasma structure.

The total reflected power can be calculated as

$$R = |\Gamma|^2 \tag{16a}$$

$$R_{\mathrm{dB}} = 10 \times \log(R) \tag{16b}$$

It is apparent that the reflected power (R) from the structure depends on plasma thickness (d), plasma electron density (N_e), plasma collision frequency (V_{en}), and incident wave frequency (ω). This makes the dispersive lossy plasma effective in controlling the overall reflections from the structure.

3.2 Four-Layered Structure

In this section, four-layered structure consisting of plastic envelope (region 1), plasma layer (region 2), RAM (region 3), and conductor (region 4) is considered. Both homogeneous and inhomogeneous unmagnetized plasma are considered.

3.2.1 RAM Covered by Homogeneous Plasma

Figure 2 shows the four-layered structure, in which EM wave propagating in—z direction is incident. The EM wave will be reflected at each interface (air–plastic, plastic–plasma, plasma–RAM, RAM–conductor). The reflections from different layers of the structure depend on the consecutive parameters of each layer.

Fig. 2 Four-layered
structure: plasma with RAM

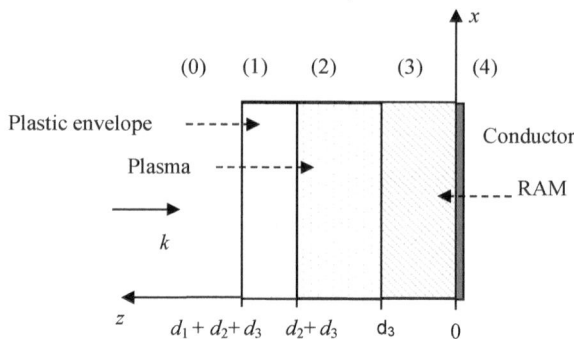

The parameters Z_0, Z_1, Z_2, Z_3, k_0, k_1, k_2, k_3, μ_0, μ_1, μ_2, μ_3, ε_0, ε_1, ε_2, ε_3 are the impedance, wave number, permeability, and permittivity in air (0th), 1st, 2nd, and 3rd layer, respectively. d_1, d_2, and d_3 are the thickness of 1st, 2nd and 3rd layer, respectively.

In 0th layer (air)

$$Z_0 = Z_0 (\text{free space impedance}) \tag{17a}$$

$$k_0 = \frac{2\pi f \left(\sqrt{\mu_0 \varepsilon_0}\right)}{c} \tag{17b}$$

The electric and magnetic fields in air are expressed as

$$E_x = E_0^+ \exp(jk_0 z) + E_0^- \exp(-jk_0 z) \tag{18a}$$

$$H_y = \frac{E_0^+ \exp(jk_0 z) - E_0^- \exp(-jk_0 z)}{Z_0} \tag{18b}$$

In the first layer (plastic envelope), the impedance is given by

$$Z_1 = \sqrt{\frac{\mu_1}{\varepsilon_1}} Z_0 \tag{19a}$$

$$k_1 = \frac{2\pi f \left(\sqrt{\mu_1 \varepsilon_1}\right)}{c} \tag{19b}$$

The electric and magnetic fields in plastic envelope are expressed as

$$E_x = E_1^+ \exp(jk_1 z) + E_1^- \exp(-jk_1 z) \tag{20a}$$

$$H_y = \frac{E_1^+ \exp(jk_1 z) - E_1^- \exp(-jk_1 z)}{Z_1} \tag{20b}$$

For second layer (plasma), the impedance is given by

$$Z_2 = \sqrt{\frac{\mu_2}{\varepsilon_2}} Z_0 \tag{21a}$$

$$k_2 = \frac{2\pi f\left(\sqrt{\mu_2 \varepsilon_2}\right)}{c} \tag{21b}$$

The electric and magnetic fields in plasma layer are expressed as

$$E_x = E_2^+ \exp(jk_2 z) + E_2^- \exp(-jk_2 z) \tag{22a}$$

$$H_y = \frac{E_2^+ \exp(jk_2 z) - E_2^- \exp(-jk_2 z)}{Z_2} \tag{22b}$$

In third layer (RAM), the impedance is expressed as

$$Z_3 = \sqrt{\frac{\mu_3}{\varepsilon_3}} Z_0 \tag{23a}$$

$$k_3 = \frac{2\pi f\left(\sqrt{\mu_3 \varepsilon_3}\right)}{c} \tag{23b}$$

The electric and magnetic fields in RAM layer are given by

$$E_x = E_3^+ \exp(jk_3 z) + E_3^- \exp(-jk_3 z) \tag{24a}$$

$$H_y = \frac{E_3^+ \exp(jk_3 z) - E_3^- \exp(-jk_3 z)}{Z_3} \tag{24b}$$

The fourth layer is conductor which totally reflects the incident wave. Thus, at $z = d_1 + d_2 + d_3$, $E_3^+ = -E_3^-$.
The electric and magnetic fields in layer 3 are expressed as

$$E_x = E_3^+ \exp(jk_3 z) - E_3^- \exp(-jk_3 z) \tag{25a}$$

$$H_y = \frac{E_3^+ \exp(jk_3 z) + E_3^- \exp(-jk_3 z)}{Z_3} \tag{25b}$$

Case 1: Two interfaces: Here only air–plastic and plastic–plasma interfaces are considered. The plasma end is assumed to be infinite.

In order to obtain the reflection coefficient, the origin of the coordinate system is shifted to left by the value of $d_2 + d_3$.

The reflection at the plastic–plasma interface is given by

$$\Gamma_{12} = \frac{Z_2 - Z_1}{Z_2 + Z_1}$$

The wave impedance in the 1st layer can be written as

$$
\begin{aligned}
Z(z) &= \frac{E_x(1)}{H_y(1)} = \frac{E_1^+ \exp(jk_1 z) + \Gamma_{12} E_1^+ \exp(-jk_1 z)}{E_1^+ \exp(jk_1 z) - \Gamma_{12} E_1^+ \exp(-jk_1 z)} \times Z_1 \\
&= \frac{\exp(jk_1 z) + \frac{Z_2 - Z_1}{Z_2 + Z_1} \exp(-jk_1 z)}{\exp(jk_1 z) - \frac{Z_2 - Z_1}{Z_2 + Z_1} \exp(-jk_1 z)} \times Z_1 \\
&= \frac{Z_2 \cosh jk_1 z + Z_1 \sinh jk_1 z}{Z_2 \sinh jk_1 z + Z_1 \cosh jk_1 z} \times Z_1
\end{aligned}
\tag{26}
$$

Applying boundary condition at air–plastic interface gives

$$E_x(0) = E_1^+; \quad z = d_1 \tag{27a}$$

$$H_y(0) = H_1^+; \quad z = d_1 \tag{27b}$$

Equations (27a) and (27b) can be written as

$$E_0^+ + E_0^- = E_1^+ \tag{28a}$$

$$\frac{E_0^+ - E_0^-}{Z_0} = \frac{E_1^+}{Z(d_1)} \tag{28b}$$

Dividing (28a) by (28b) gives

$$Z_0 \frac{E_0^+ + E_0^-}{E_0^+ - E_0^-} = Z(d_1) \tag{29}$$

$$\frac{E_0^-}{E_0^+} = \frac{Z(d_1) - Z_0}{Z(d_1) + Z_0} \tag{30}$$

Here $Z(d_1)$ is the impedance beyond the first interface and can be determined from (26) as follows:

$$Z(d_1) = \frac{Z_2 \cosh jk_1 d_1 + Z_1 \sinh jk_1 d_1}{Z_2 \sinh jk_1 d_1 + Z_1 \cosh jk_1 d_1} \times Z_1 \tag{31}$$

At first interface, (air–plastic) $Z(d_1)$ can be taken as input impedance, Z_{in1}. Thus, (30) can be generalized as

$$\frac{E_0^-}{E_0^+} = \frac{Z_{in1} - Z_0}{Z_{in1} + Z_0} = \Gamma \tag{32}$$

where Γ is the reflection coefficient of the two-interface case (Fig. 3). Using (16a) and (16b), the reflected power can be determined.

Case 2: Three interfaces: Now, the RAM layer is included in two-interface case. The three-interface structure is shown in Fig. 4.

It is noted that Z_{in1} is the impedance beyond $d_1 + d_2$. Further, Z_2 in (31) is the impedance beyond second interface and it can be denoted as Z_{in2}. Equation (31) can be modified by using Z_{in2} as follows:

$$Z_{in1} = \frac{Z_{in2} \cosh jk_1 d_1 + Z_1 \sinh jk_1 d_1}{Z_{in2} \sinh jk_1 d_1 + Z_1 \cosh jk_1 d_1} \times Z_1 \tag{33a}$$

Fig. 3 Two interface case

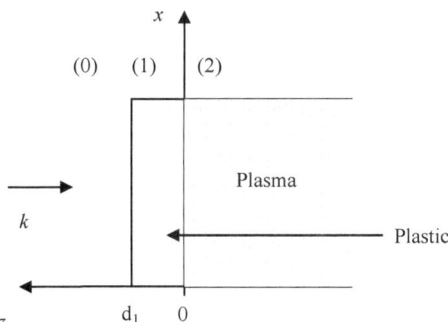

Fig. 4 Three interface case

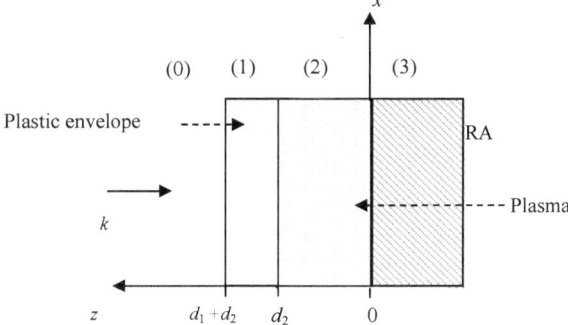

where

$$Z_{\text{in2}} = \frac{Z_3 \cosh jk_2 d_2 + Z_2 \sinh jk_2 d_2}{Z_3 \sinh jk_2 d_2 + Z_2 \cosh jk_2 d_2} \times Z_2 \tag{33b}$$

It may be observed that the impedance beyond the first interface (air–plastic) is modified by the impedance beyond the second interface (plastic–plasma).

Case 3: Four interfaces: Next, the fourth layer of conductor is added after RAM. The four-interface structure is shown in Fig. 5.

Here Z_{in2} is the impedance beyond $d_2 + d_3$. Due to the presence of third layer boundary $Z_3 = Z_{\text{in3}}$ (impedance beyond second interface, d_3). Thus, from (33b), one has

$$Z_{\text{in2}} = \frac{Z_{\text{in3}} \cosh jk_2 d_2 + Z_2 \sinh jk_2 d_2}{Z_{\text{in3}} \sinh jk_2 d_2 + Z_2 \cosh jk_2 d_2} \times Z_2 \tag{33c}$$

where

$$Z_{\text{in3}} = \frac{Z_4 \cosh jk_3 d_3 + Z_3 \sinh jk_3 d_3}{Z_4 \sinh jk_3 d_3 + Z_3 \cosh jk_3 d_3} \times Z_3 \tag{33d}$$

By transforming impedance from right side of the structure to the left side, reflection from the structure can be determined. The structure can be modeled, as shown in Fig. 6. The fourth layer is conductor so $Z_4 = 0$. Equation (33d) reduces to

$$Z_{\text{in3}} = \frac{Z_3 \sinh jk_3 d_3}{Z_3 \cosh jk_3 d_3} \times Z_3 = Z_3 \tanh jk_3 d_3 = Z_3 \delta_3 \tag{34}$$

where $\delta_3 = \tanh jk_3 d_3$

Fig. 5 Four interface case

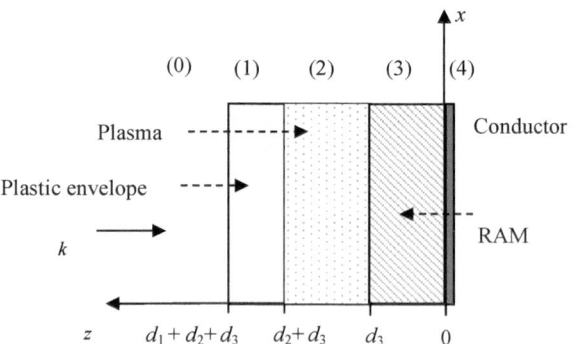

Fig. 6 Four-layered model

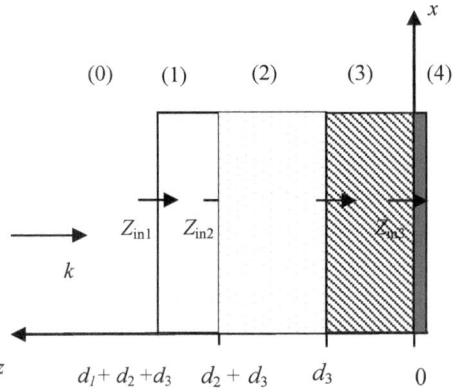

Using (34), the Eq. (33c) reduces to

$$Z_{in2} = \frac{Z_3\delta_3\cosh jk_2d_2 + Z_2\sinh jk_2d_2}{Z_3\delta_3\sinh jk_2d_2 + Z_2\cosh jk_2d_2} \times Z_2$$
$$= \frac{Z_3\delta_3 + Z_2\tanh jk_2d_2}{Z_3\delta_3\tanh jk_2d_2 + Z_2} \times Z_2 = \frac{Z_3\delta_3 + Z_2\delta_2}{Z_3\delta_3\delta_2 + Z_2} \times Z_2 \tag{35}$$

where $\delta_2 = \tanh jk_2d_2$.

Likewise, using (35), the Eq. (33a) reduces to

$$Z_{in1} = \frac{\frac{Z_3\delta_3 + Z_2\delta_2}{Z_3\delta_3\delta_2 + Z_2} \times Z_2 \times \cosh jk_1d_1 + Z_1\sinh jk_1d_1}{\frac{Z_3\delta_3 + Z_2\delta_2}{Z_3\delta_3\delta_2 + Z_2} \times Z_2 \times \sinh jk_1d_1 + Z_1\cosh jk_1d_1} \times Z_1$$
$$= \frac{\frac{Z_3\delta_3 + Z_2\delta_2}{Z_3\delta_3\delta_2 + Z_2} \times Z_2 + Z_1\tanh jk_1d_1}{\frac{Z_3\delta_3 + Z_2\delta_2}{Z_3\delta_3\delta_2 + Z_2} \times Z_2 \times \tanh jk_1d_1 + Z_1} \times Z_1$$
$$= \frac{\frac{Z_3\delta_3 + Z_2\delta_2}{Z_3\delta_3\delta_2 + Z_2} \times Z_2 + Z_1\delta_1}{\frac{Z_3\delta_3 + Z_2\delta_2}{Z_3\delta_3\delta_2 + Z_2} \times Z_2 \times \delta_1 + Z_1} \times Z_1 \tag{36}$$

where $\delta_1 = \tanh jk_1d_1$.

Equation (36) can be rearranged as

$$Z_{in1} = \frac{Z_2Z_3\delta_3 + Z_2^2\delta_2 + Z_1Z_3\delta_1\delta_2\delta_3 + Z_1Z_2\delta_1}{Z_2Z_3\delta_1\delta_3 + Z_2^2\delta_1\delta_2 + Z_1Z_3\delta_2\delta_3 + Z_1Z_2} \times Z_1 \tag{37}$$

Substituting (37) in (32), the reflection coefficient of the four-layered structure can be determined. The corresponding reflected power, R can be obtained from (16a) and (16b).

3.2.2 RAM Covered by Inhomogeneous Plasma

As mentioned in previous section, the electron density gradient in the plasma alters the overall reflection behavior of the structure. Here the inhomogeneous plasma is analyzed by dividing the plasma medium into equal sublayers. Each sublayer has same thickness and collision frequency, but with distinct electron density. The overall electron density of the layer follows specific type of density profile. Distinct electron density in each sublayer makes the dielectric constant of the sublayer different from other sublayers.

The inhomogeneous plasma can be treated as multilayered dielectric medium and hence can be analyzed using impedance transformation method. In each electron density profile, the electron density increases from the air–plasma interface. In other words, the electron density is minimum at the air–plasma interface and increases gradually to its maximum at the plasma–conductor interface. In this document, three types of density profiles are taken into account. These are linear, parabolic, and exponential.

Linear electron density profile: The expression for linear electron density profile of plasma is given by (Gruel and Oncu 2009)

$$N_e(z) = \frac{N_e z}{d} \tag{38}$$

where d is the plasma thickness and N_e is the maximum electron density in the plasma. In order to solve the bounded inhomogeneous plasma problem as the multilayered dielectric medium using impedance transformation method, layering of the plasma with specified density profile is required. Figure 7 shows the dis-

Fig. 7 Discretization of linear electron density profile

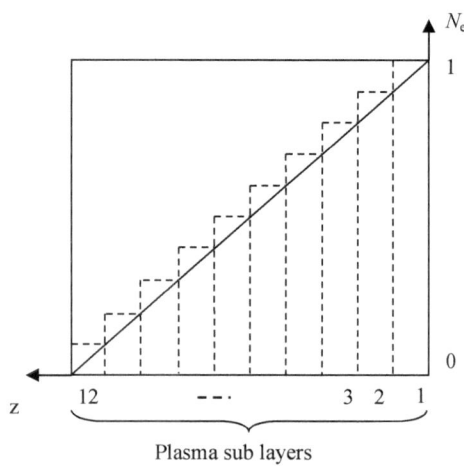

Plasma sub layers

cretization of the linear density profile. Here the plasma layer is divided into twelve sublayers. The electron density in the nth plasma sublayer is expressed as

$$N_e(n) = \frac{N_e(t - (n-1))}{t} \tag{39}$$

where $n = 1, 2, 3, \ldots, 12$ and t is the number of sublayers in the plasma.

Parabolic electron density profile: The parabolic electron density profile of the plasma is expressed as (Tang et al. 2003)

$$N_e(z) = N_e \exp\frac{2}{3}\left(\frac{2Z}{d} - 1\right); \quad \text{for} \quad Z < \frac{d}{2} \tag{40a}$$

$$N_e(z) = N_e \exp\frac{2}{3}\left(1 - \frac{2Z}{d}\right); \quad \text{for} \quad \frac{d}{2} < z < d \tag{40b}$$

where d is the plasma thickness and N_e is the maximum electron density in the plasma. Figure 8 shows the discretization of the parabolic density profile.

The electron density in the nth plasma sublayer can be written as

$$N_e(z) = N_e \exp\frac{2}{3}\left(\frac{2(t - (n-1))}{t} - 1\right) \quad \text{for} \quad Z < \frac{d}{2} \tag{41a}$$

$$N_e(z) = N_e \exp\frac{2}{3}\left(1 - \frac{2(t - (n-1))}{t}\right) \quad \text{for} \quad \frac{d}{2} < z < d \tag{41b}$$

where t is the number of sublayers in the plasma.

Fig. 8 Discretization of parabolic electron density profile

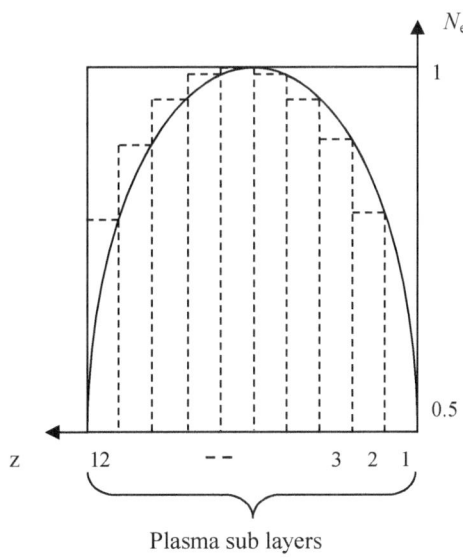

Plasma sub layers

Fig. 9 Discretization of
exponential electron density
profile

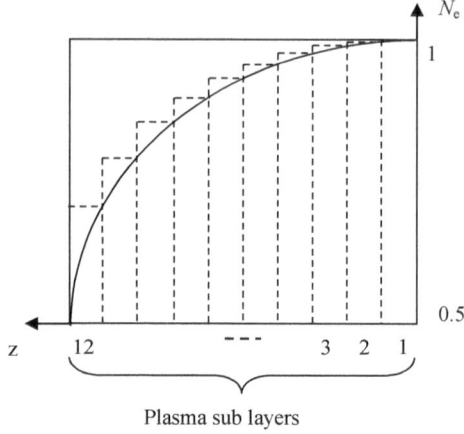

Plasma sub layers

Exponential electron density profile: This density profile of the plasma follows
the expression given by (Tang et al. 2003)

$$N_e(z) = N_e \exp\frac{2}{3}\left(\frac{Z}{d} - 1\right) \tag{42}$$

The electron density in the nth plasma sublayer is expressed as

$$N_e(z) = N_e \exp\frac{2}{3}\left(\frac{(t - (n-1))}{t} - 1\right) \tag{43}$$

where t is the number of sublayers in the plasma. Figure 9 shows the discretization
of the exponential density profile.

4   Results and Discussion

In this section, the computed results for the reflection characteristics of bounded
plasma are presented. The reflection behavior of the multilayered structure with
plasma, RAM, and conductor is analyzed based on the parameters, viz. plasma
thickness (d), plasma collision frequency (V_{en}), and plasma electron density (N_e).

The inhomogeneity in the plasma alters the reflection behavior of the structure.
This effect of inhomogeneity in the plasma layer is studied by considering different
electron density profiles in the plasma medium.

4.1 Two-Layered Structure: Bounded Plasma

Here, structure considered consists of plasma and conductor.

4.1.1 Effect of Plasma Thickness

As discussed, the plasma thickness is the parameter that gives rise to cavity resonance effect in the plasma medium. Thus, it is required to optimize the plasma
thickness for a given electron density, collision frequency, and incident wave frequency. This plasma thickness optimization is required to achieve maximum EM
wave absorption. Figures 10 and 11 present the dependence of reflected power on
the incident wave frequency. The computed results are validated against the results
available in open domain (Yuan et al. 2010).

The results are shown for different plasma thicknesses. It may be observed from
Fig. 10 that for f_p = 2 GHz, V_{en} = 5 GHz, the structure shows maximum absorption for d = 8.6 cm. On the other hand for f_p = 8 GHz, V_{en} = 10 GHz (Fig. 11),
maximum absorption takes place at d = 10 cm. In order to have more clarity,
another case of two-layered structure with N_e = 2 × 10^{17} m^{-3}, V_{en} = 10 GHz is
considered. The variation of reflected power as a function of incident wave frequency is shown in Fig. 12. It is apparent that maximum absorption corresponds to
d = 4.4 cm.

The optimum plasma thickness for maximum EM wave absorption depends on
the collision frequency and the plasma electron density. This may be seen in
Fig. 13, where reflected power is computed for different collision frequency,
plasma thickness, and the plasma electron density. The maximum absorption occurs
at different incident frequencies.

Table 1 presents the optimum plasma thickness for different collision frequency
and electron density. The increase in collision frequency and electron density
changes the optimum plasma thickness.

Figure 14 shows that the reflection characteristics of plasma (N_e = 2 × 10^{17} m^{-3},
V_{en} = 10 GHz) for different plasma thicknesses.

It may be noted that the maximum EM wave absorption happens not only for
one particular plasma thickness but occurs at different plasma thicknesses (similar
to harmonics). As the plasma thickness increases the absorption level and the
resonant frequency increases. However, the bandwidth of absorption decreases with
increase in the plasma thickness (Table 2).

Figure 15 shows the reflected power for different incident wave frequency. The
computed results are validated against the results available in open domain (Yuan
et al. 2010). It may be noted that for f = 5 GHz, the reflected power decrease with
increase in the plasma thickness. Another important point to be noted is that this
behavior is shown when the incident wave frequency is same as plasma collision
frequency. This is not true for frequencies such as f = 1 GHz and f = 3 GHz. At
f = 1 GHz, the incident wave frequency is less than the plasma frequency

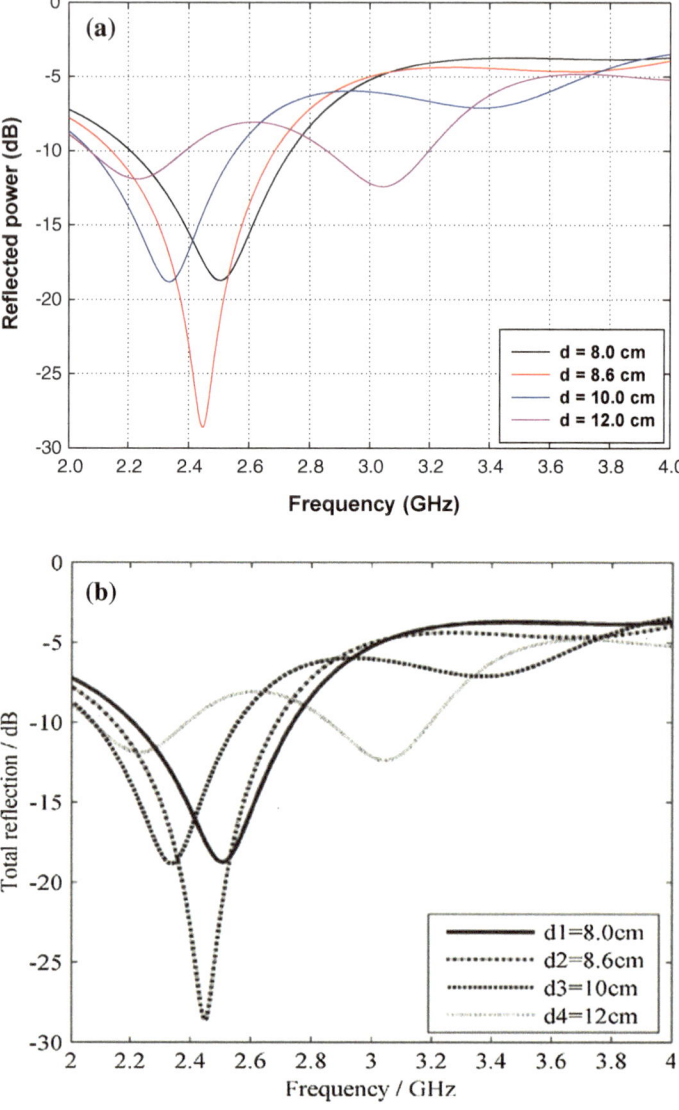

Fig. 10 Dependence of reflected power on incident wave frequency $f_p = 2$ GHz, $V_{en} = 5$ GHz. **a** Computed. **b** Reference Yuan et al. (2010)

(~ 2 GHz corresponding to $N_e = 5 \times 10^{16}$ m^{-3}), the reflected power remains constant for different plasma thickness. The small decrease in reflected power may be due to the collisions within the plasma layer. For $f = 3$ GHz, the reflected power oscillates and then becomes constant with increase in plasma thickness.

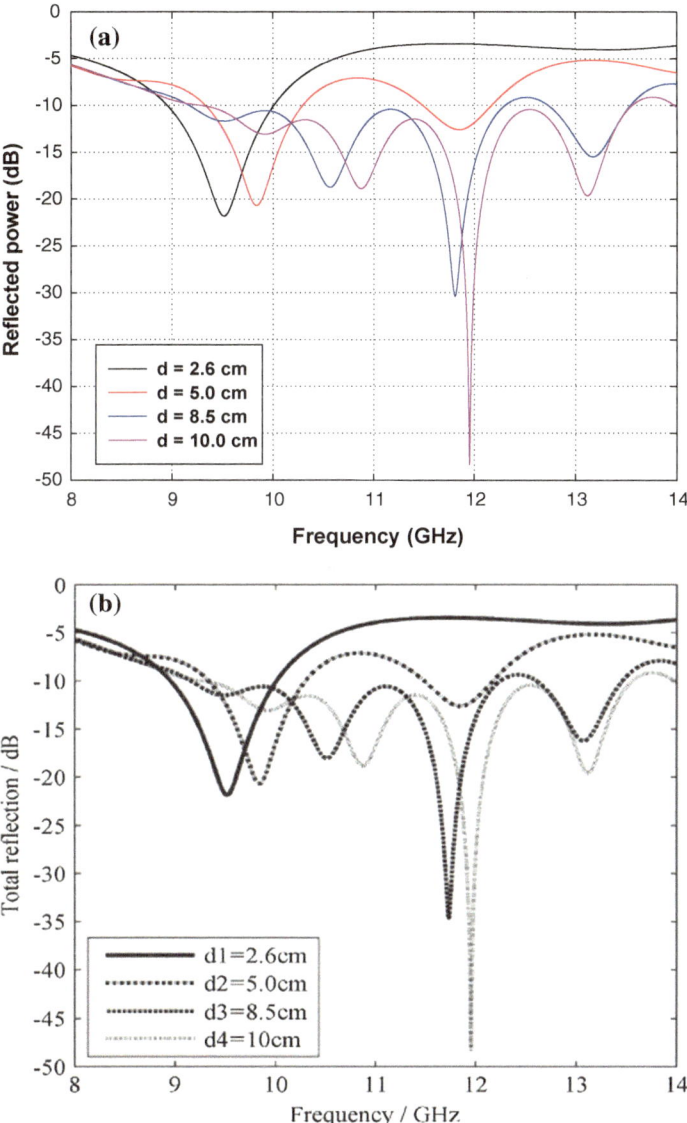

Fig. 11 Dependence of reflected power on incident wave frequency $f_p = 8$ GHz, $V_{en} = 10$ GHz.
a Computed. **b** Reference Yuan et al. (2010)

When incident wave frequency is close to but higher than plasma frequency, the absorption takes place for thin layer of plasma. On the other hand, for incident wave frequency higher than plasma frequency, e.g., $f = 5$ GHz, the absorption is less for thin plasma layer. For greater absorption, thicker plasma is required.

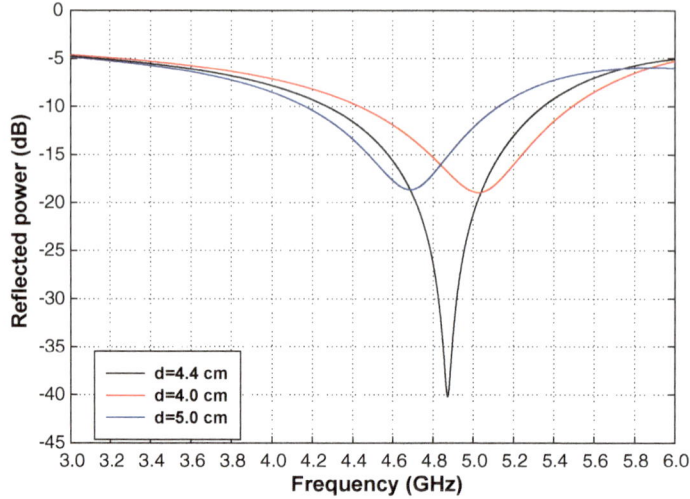

Fig. 12 Reflected power versus incident wave frequency $N_e = 2 \times 10^{17}$ m^{-3}, $V_{en} = 10$ GHz

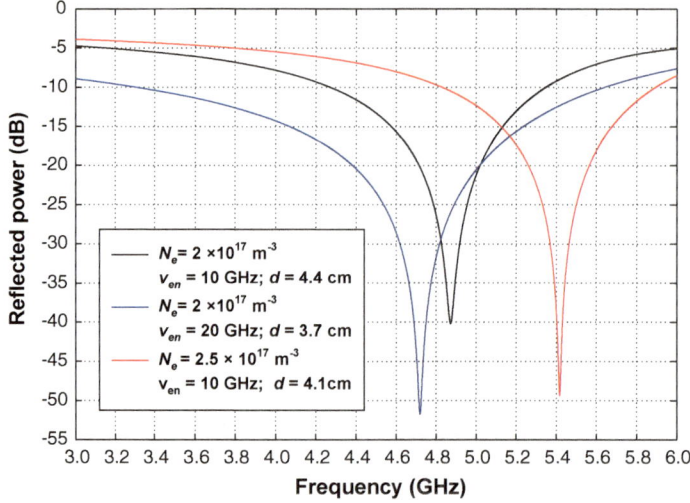

Fig. 13 Dependence of maximum absorption on plasma parameters

In Fig. 16, the reflected power of the structure is shown for different plasma collision frequencies. There is no particular trend in variation of reflected power. This may be due to the combined effect of collisional absorption and cavity resonance effect. However, one can infer that for plasma collision frequency greater than the incident frequency ($V_{en} = 10$ GHz, $f = 3$ GHz), the absorption takes place for thin plasma. For very high collision frequency ($V_{en} = 100$ GHz), the absorption

Table 1 Optimized plasma thickness for different plasma parameters for two-layered structure

Plasma parameters	Optimized plasma thickness (cm)
f_p = 2 GHz, V_{en} = 5 GHz	8.6
f_p = 8 GHz, V_{en} = 10 GHz	10.0
N_e = 2 × 10^{17} m^{-3}, V_{en} = 10 GHz	4.4
N_e = 2 × 10^{17} m^{-3}, V_{en} = 20 GHz	3.7
N_e = 2.5 × 10^{17} m^{-3}, V_{en} = 10 GHz	4.1

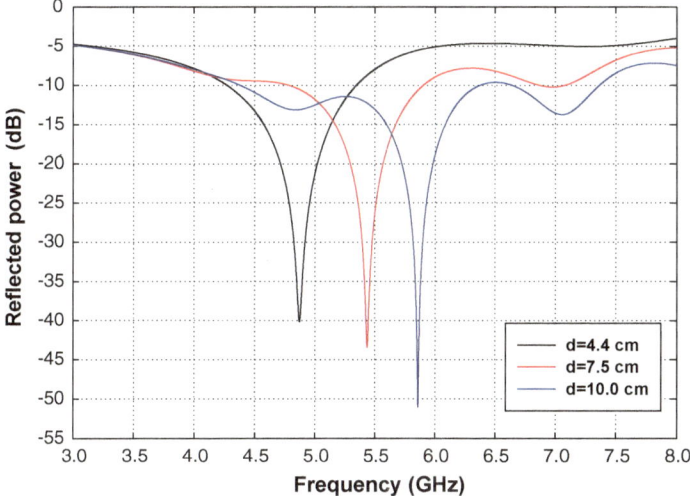

Fig. 14 Dependence of absorption on plasma thickness N_e = 2 × 10^{17} m^{-3}, V_{en} = 10 GHz

Table 2 Absorption behavior of plasma for different thickness

Plasma thickness (cm)	Maximum absorption (dB)	Resonant frequency (GHz)	BW (GHz)
4.4	40.17	4.87	0.3
7.5	43.45	5.44	0.27
10.0	51.02	5.88	0.26

takes place at higher plasma thickness. The computed results are validated against the results available in open domain (Yuan et al. 2010).

Next, the variation of reflected power with plasma thickness for different ratio of plasma frequency and incident frequency is studied (Fig. 17). The incident frequency is kept constant at 3 GHz. The computed results are validated against the results available in open domain (Yuan et al. 2010). It can be inferred that for greater absorption, the plasma frequency should be greater than the incident frequency.

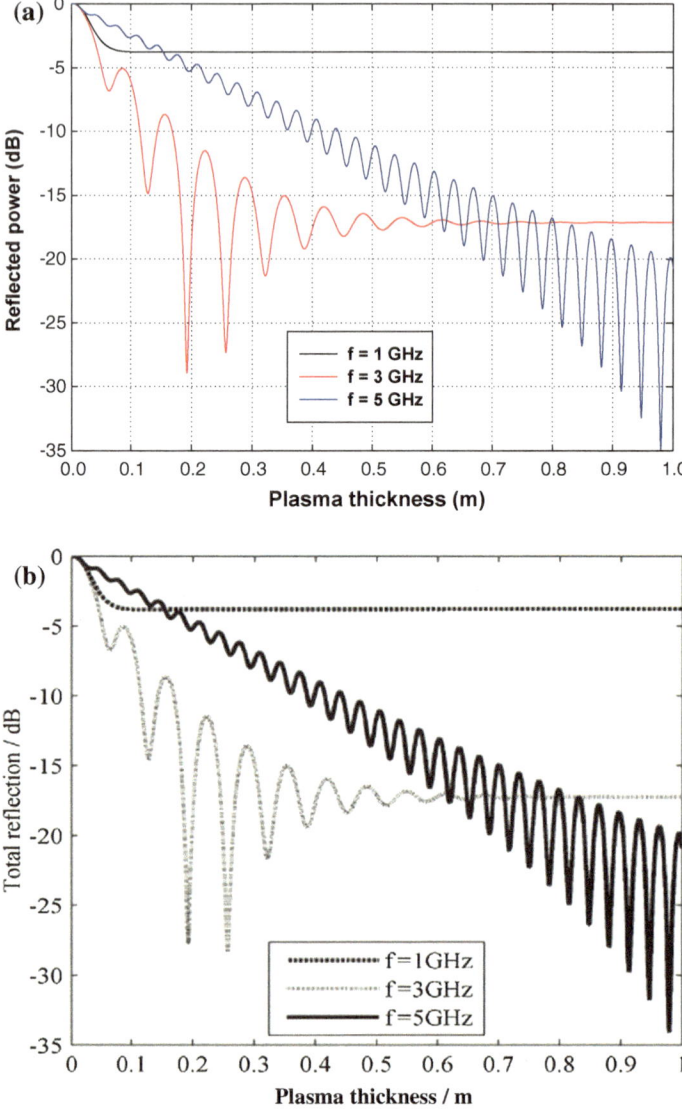

Fig. 15 Reflected power for different incident wave frequency; $N_e = 5 \times 10^{16}$ m^{-3}, $V_{en} = 5$ GHz. **a** Computed. **b** Reference Yuan et al. (2010)

In Fig. 18, the reflected power of two-layered conductor-backed plasma structure is shown for different incident frequencies. It may be observed that when incident frequency is less than plasma frequency of 2 GHz with $N_e = 5 \times 10^{16}$ m^{-3}, the absorption in plasma is very less, and does not vary with plasma thickness. If the incident frequency is greater but close to the plasma frequency ($f = 3$ GHz), the absorption takes place only for thin plasma. For $f = 5$ GHz, which is greater than the

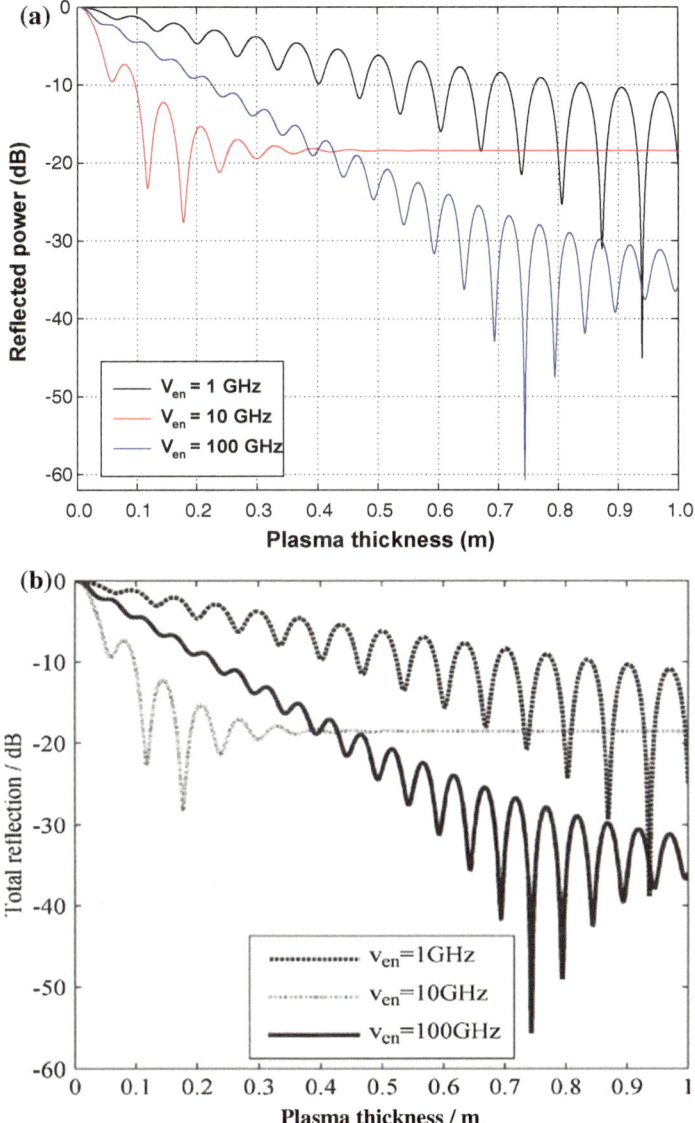

Fig. 16 Dependence of reflected power on plasma collision frequency $N_e = 5 \times 10^{16}$ m^{-3}, $f = 3$ GHz. **a** Computed. **b** Reference Yuan et al. (2010)

plasma frequency, the absorption becomes more as the plasma thickness is increased. If further the incident frequency is increased, i.e., $f = 10$ GHz, the absorption gets shifted toward very high plasma thickness.

Figure 19 presents the role of plasma electron density on the absorption behavior of conductor-backed plasma structure. As one knows that the plasma frequency is

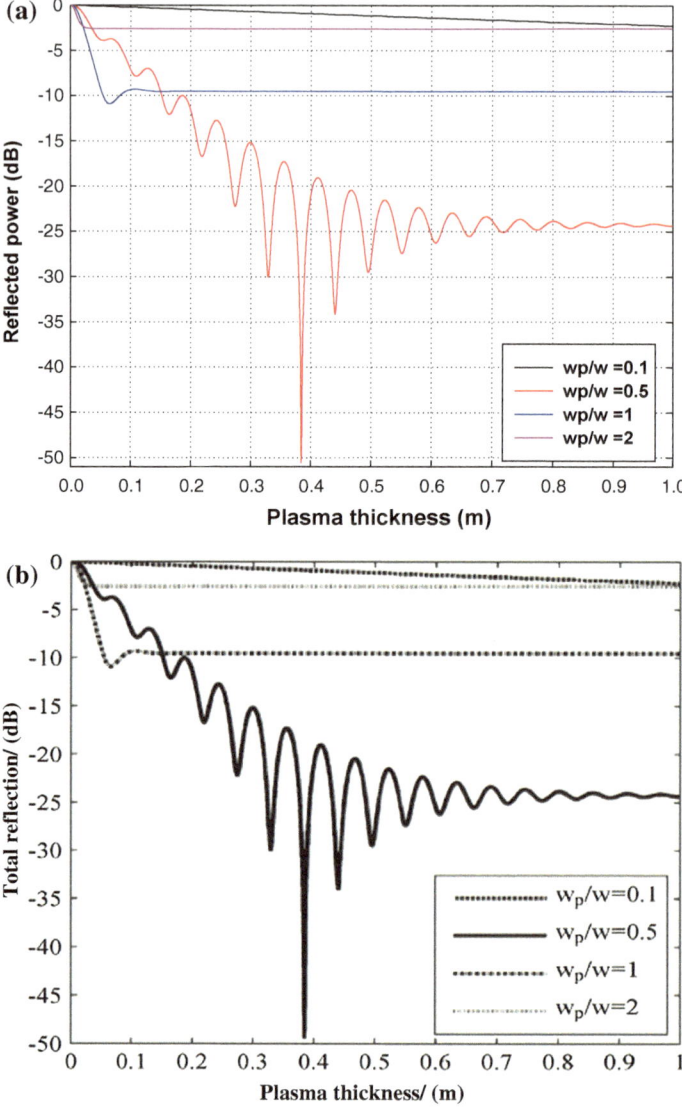

Fig. 17 Variation of reflected power with plasma thickness for different ratio of plasma frequency and incident frequency; $f = 3$ GHz, $V_{en} = 10$ GHz. **a** Computed. **b** Reference Yuan et al. (2010)

dependent on plasma electron density, for $N_e = 5 \times 10^{17}$ m^{-3}, the plasma frequency is 6.35 GHz. This frequency is greater than the incident frequency, and hence the reflected power is more and does not vary with plasma thickness.

This is not the case for other electron density considered because the corresponding plasma frequencies are less than the incident frequency. Here one may note that maximum absorption take place for $N_e = 1 \times 10^{17}$ m^{-3}, $\omega_p = 2.84$ GHz

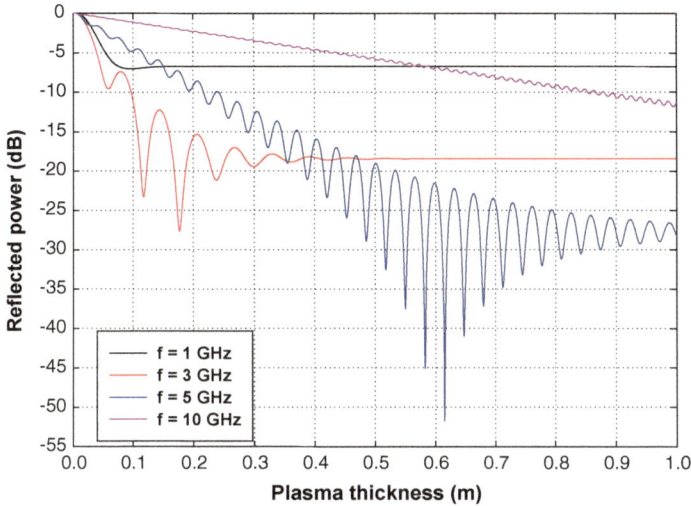

Fig. 18 Reflected power of two-layered conductor-backed plasma structure for different incident frequency; $N_e = 5 \times 10^{16}$ m^{-3}, $V_{en} = 10$ GHz

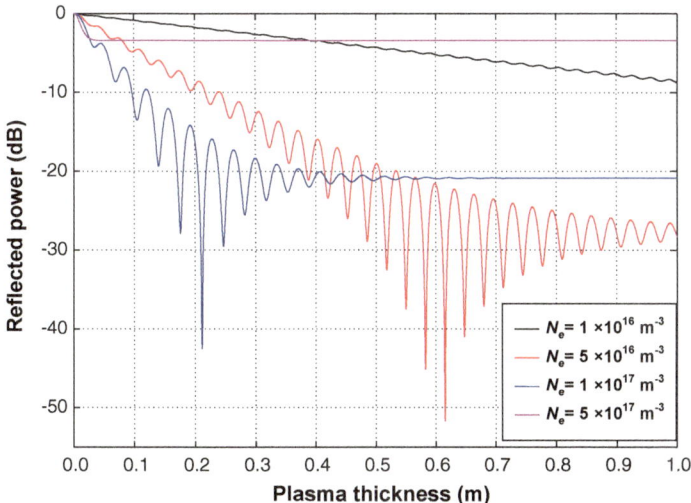

Fig. 19 Variation of reflected power with plasma thickness for different plasma electron density; $f = 5$ GHz, $V_{en} = 10$ GHz

for thin plasma, whereas for $N_e = 5 \times 10^{16}$ m^{-3}, $\omega_p = 2$ GHz, absorption happens for thicker plasma.

Figure 20 presents the variation of reflected power of two-layered metal-backed plasma structure for different collision frequencies. For higher collision frequencies

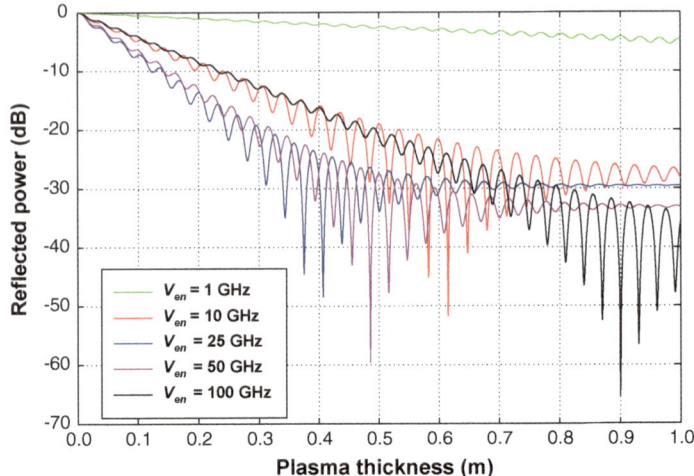

Fig. 20 Variation of reflected power for two-layered metal-backed plasma structure with plasma thickness for different collision frequencies $N_e = 5 \times 10^{16}$ m^{-3}, $f = 5$ GHz

($V_{en} = 10$ GHz, 25 GHz, 50 GHz) as compared to incident frequency of 5 GHz, the absorption take place for plasma thickness up to 0.5 m. However for very high collision frequency of 100 GHz, the absorption shifts toward plasma thickness greater than 0.5 m.

One may infer that the cavity resonance effect shows ripples in the reflection characteristics of the bounded plasma. The effect of cavity resonance vanishes for thick plasma.

From (14), $Z(d) = Z_2 \tanh j k_2 d$.

The impedance saturates due to unity value of tanh() term for high values of $k_2 d$. The oscillatory behavior of impedance and that of the reflection coefficient vanishes. Further, the reflected power decreases to minimum and then saturates as the plasma thickness increases. The maximum absorption, saturated reflected power, and the corresponding plasma thickness depend on the electron density, collision frequency, and the incident wave frequency, respectively.

The cavity resonance effects on the reflection characteristics of bounded plasma for different plasma parameters are summarized in Table 3.

4.1.2 Effect of Collision Frequency

The absorption in the plasma is due to collisions within the plasma. For bounded plasma, the absorption is the combined effect of collisional absorption and cavity resonance effect. Figure 21 shows the effect of collision frequency on the reflection characteristics of the two-layered conductor-backed plasma. As the collision frequency increases the reflection characteristics slightly shift toward low frequency

Table 3 Effect of plasma parameters on the reflection characteristics of two-layered metal-backed plasma structure

Plasma parameter	Performance
Electron density (N_e) Plasma frequency (ω_p) Incident wave frequency (ω)	$\omega < \omega_p$: No variation in reflection characteristics as the plasma thickness increases $\omega > \omega_p$: Ripples arises in the reflection characteristics As the difference ($\omega - \omega_p$) increases, • Maximum absorption and the corresponding plasma thickness increases • Saturated reflection loss and the corresponding plasma thickness increases
Collision frequency (V_{en})	Maximum reflection loss: Irregular variation (due to the combined effect of collisional absorption and cavity resonance effect)
	As V_{en} approaches ω: (V_{en} less than or close to ω) Saturated reflection loss remains same Plasma thicknesses at which maximum absorption and saturated reflection loss occurs decrease (variation becomes small when V_{en} is close to ω)
	When V_{en} increased beyond ω: ($V_{en} \gg \omega$) Saturated reflection loss increases Plasma thicknesses at which, maximum absorption and saturated reflection loss occurs increases

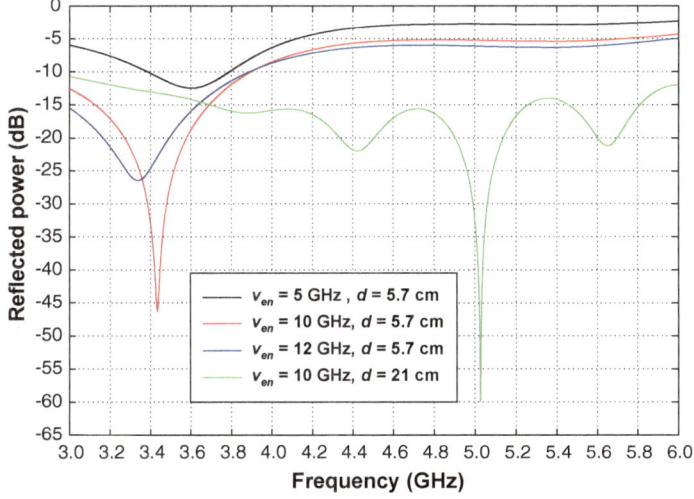

Fig. 21 Dependence of reflected power of two-layered conductor-backed plasma on collision frequency and the plasma thickness, $N_e = 1 \times 10^{17}\,\mathrm{m}^{-3}$

region. At particular collision frequency the absorption is maximum, depending upon the plasma thickness.

Further, it is apparent that the reflected power decreases with the increases in collision frequency except for the resonant absorption band. This is because in resonant absorption band, EM wave absorption is not only due to the collisions but also due to the cavity resonance effect. The multiple reflections in the plasma layer are modified by the collisions within the plasma layer and at interfaces, giving rise to overall reflection behavior of the structure.

At optimum plasma thickness and collision frequency, the phases will be modified toward maximum EM wave absorption. The location of the resonant absorption band is determined by the electron density of the plasma. This electron density in turn controls the plasma frequency. In the low frequency region, the absorption is mainly due to the collisions within the plasma. At high incident frequency end ($\omega - \omega_p$ is high) thick plasma is required for effective absorption. This is due to cavity resonance effect. When the incident wave frequency is comparable with the plasma frequency, thin plasma has maximum absorption. Thus one can infer that at high incident frequencies ($\omega - \omega_p$ is high), thick plasma slab will have better absorption.

4.1.3 Effect of Electron Density

The plasma electron density determines the plasma frequency. As the electron density increases, the reflection characteristics shift toward the high incident frequency. The reflection characteristics of two-layered conductor-backed plasma for different electron density with $V_{en} = 10$ GHz and $d = 5.7$ cm are shown in Fig. 22. It may be observed that $N_e = 1 \times 10^{17}$ m^{-3} have maximum absorption at $V_{en} = 10$ GHz and $d = 5.7$ cm. In order to achieve maximum EM wave absorption, optimization of the plasma thickness and the collision frequency is required. If electron density is increased to $N_e = 2 \times 10^{17}$ m^{-3}, the optimized plasma thickness and collision frequency for maximum absorption are $V_{en} = 35$ GHz and $d = 6$ cm.

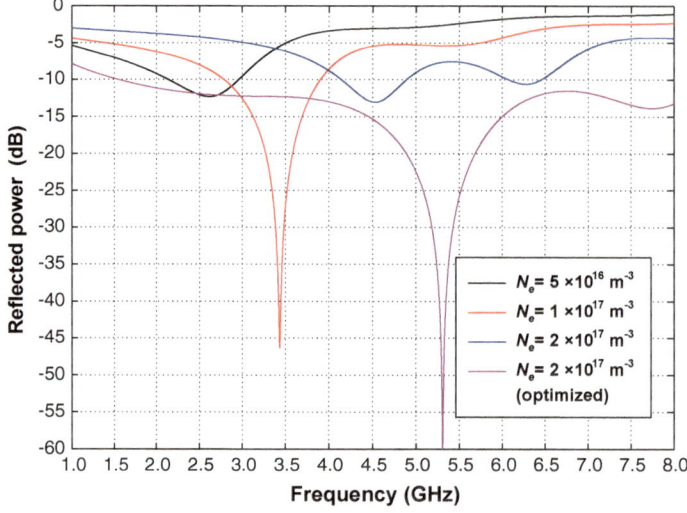

Fig. 22 Reflected power of two-layered conductor-backed plasma for different plasma electron density

4.2 Four-Layered Structure: RAM Covered by Plasma

In this section a four-layered structure consisting of plastic, plasma, RAM and conductor is considered. The reflected powers of the bounded plasma, with and without RAM layer, are different. This is mainly due to significant absorption of propagating EM wave in RAM layer. For an efficient absorption, matching between RAM parameters and plasma parameters are required. Further, choosing optimum plasma parameters and RAM, a broad absorption band can be achieved. Here, two types of RAM materials are considered (Yuan et al. 2011) for the reflection behavior analysis of the structure, are given in Table 4.

4.2.1 Effect of RAM Layer

The parameters used for the simulation of four-layered structure are tabulated in Table 5. In Fig. 23, the reflected power is shown for four-layered structure (plastic–plasma–RAM–conductor), and three-layered structures (plastic–plasma–conductor), (plastic–RAM–conductor). It may be observed that the reflected power shows minima at particular incident frequency. This is especially true for four-layered plasma–RAM structure.

At low incident frequencies, the plasma with RAM follows the same trend as the plasma without RAM case. The reflection characteristics of plasma with RAM and without RAM are different when the incident wave frequency is increased beyond the plasma frequency. A four-layered plasma-RAM structure shows better absorption. The improvement in reflection characteristics is due to the EM wave absorption in plasma and RAM.

At high frequencies plasma with RAM has reduced reflected power, where the main absorption may be due to the RAM layer in the structure. In order to achieve more absorption in plasma at high frequencies ($\omega \gg \omega_p$), thicker plasma layer is required. The reflection characteristics show a resonance absorption band near the

Table 4 Relative parameters of RAM ($f_g = f \times 10^{-9}$ GHz)

Type of RAM	$\varepsilon_r(f)$	$\mu_r(f)$
Lossy dielectric	$\frac{3}{f_g^{0.778}} - i \times \frac{15}{f_g^{0.861}}$	1
Lossy magnetic	15	$\frac{306.25}{f_g^2 + 12.25} - i \times \frac{87.5 \times f_g}{f_g^2 + 12.25}$

Table 5 Layer parameters (Fig. 24)

Structure	Thickness (cm)	Parameters
Layer 1 (plastic envelope)	1	$\varepsilon_r = 1$, $\mu_r = 1$ (free space parameters)
Layer 2 (plasma)	4.9	$N_e = 5 \times 10^{17}$ m^{-3}, $\omega_p = 6.35$ GHz, $V_{en} = 30$ GHz
Layer 3 (RAM)	1	Lossy magnetic

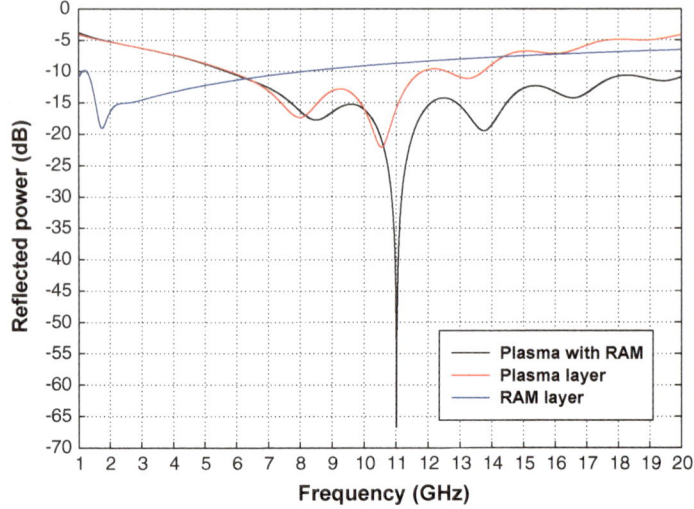

Fig. 23 Reflected power for different structures

plasma frequency. Absorption peaks in the reflection characteristics of the plasma with and without RAM shows phase shift near the resonance absorption band. This may be due to the fact the phase of the reflections get altered by the RAM layer. The resonance absorption band depends on the plasma and RAM parameters.

The resonance absorption band disappears, if type 1 (lossy dielectric) material is used for RAM with parameters given in Table 5. This emphasizes the need of optimization of parameters.

Figure 24 shows the optimization of reflected power of four-layered plasma-RAM structure. The RAM material is taken as lossy dielectric. It may be observed that the plasma thickness optimization ($d = 4.9$ cm, 4.5 cm, $V_{en} = 30$ GHz) shifts the absorption peaks toward maximum EM wave absorption at 11 GHz. However keeping plasma thickness $d = 4.5$ cm, the collision frequency is reduced to 28 GHz. This change makes the absorption even better.

In Fig. 25, the reflected power of four-layered plasma–RAM structure is shown with optimized plasma parameters. The performance is compared with three-layered structures (plastic–plasma–conductor), (plastic–RAM–conductor). It may be noted that the plasma and RAM (type 1) parameters affect the resonant absorption band. If optimum thickness is used for plasma without RAM, the reflected power from the structure can be reduced further (Fig. 26).

In Fig. 26, the plasma thickness is changed to $d = 5.4$ cm, $V_{en} = 30$ GHz. It is apparent that for this plasma thickness, the absorption shown by three-layered structure (plastic–plasma–conductor) is better than the two four-layered plasma-RAM structures. This is to emphasize the need of optimization of plasma parameters. It is not always that four-layered plasma–RAM structure will show better absorption characteristics.

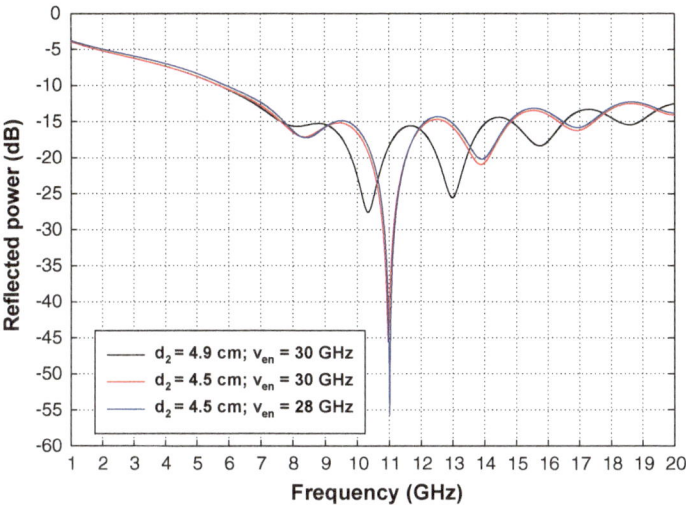

Fig. 24 Optimization of reflected power of four-layered plasma-RAM structure (type 1 RAM)

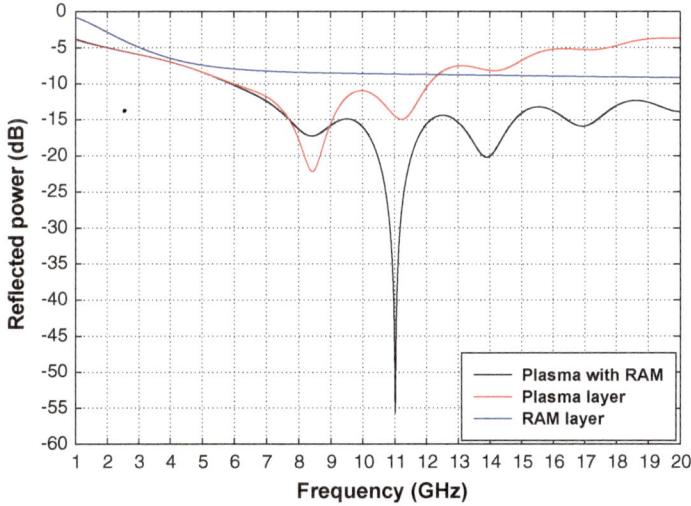

Fig. 25 Comparison of reflection behavior of four-layered and three-layered structures

However at high incident frequencies, the performance of four-layered plasma-RAM structures is better. The performance analysis of structures, coated with plasma and RAM are summarized in Table 6. It is apparent from Table 6 that the RAM coating widens the absorption band and increases the reflection loss (absorption).

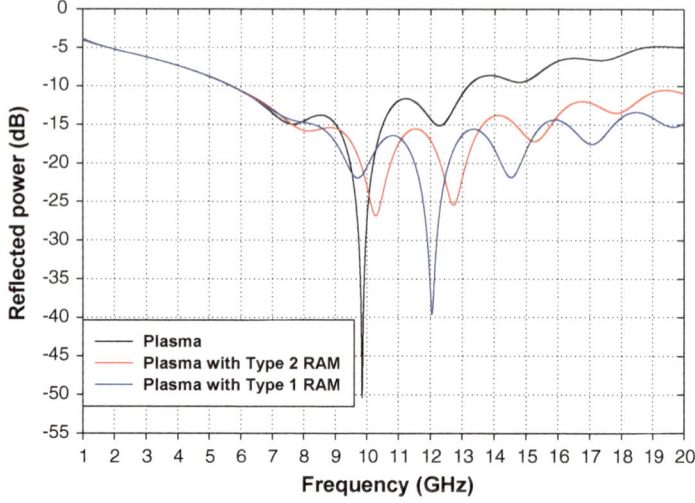

Fig. 26 Comparison of reflection behavior of three-layered plasma structure and four-layered plasma–RAM structures

Table 6 Performance of the structure ($N_e = 5 \times 10^{17}$ m^{-3}, $d_1 = 1$ cm, $d_3 = 1$ cm)

Structure	Optimized plasma parameters	Peak reflection (dB)	Resonant frequency (GHz)	BW (GHz)
Plasma without RAM	$V_{en} = 30$ GHz, $d_2 = 5.4$ cm	49.88	9.85	0.73
Plasma with RAM (type 2)	$V_{en} = 30$ GHz, $d_2 = 4.9$ cm	66.69	11.02	1.03
Plasma with RAM (type 1)	$V_{en} = 28$ GHz, $d_2 = 4.5$ cm	55.49	11.02	1.05

4.2.2 Effect of Plastic Envelope

The parameters of the plastic envelope affect the overall reflections from the structure. Figure 27 shows the effect of plastic layer parameters on the overall reflections from three-layered structure (plastic–plasma–conductor). As compared to free space, the high dielectric constant of plastic envelope creates few absorption peaks in the reflection characteristics. The location and depth of these absorption peaks depend on the constitutive parameters of plastic envelope.

4.2.3 Effect of Plasma Parameters

Plasma thickness and plasma collision frequency are the main parameters that control EM wave absorption in the plasma. It is known that the plasma electron

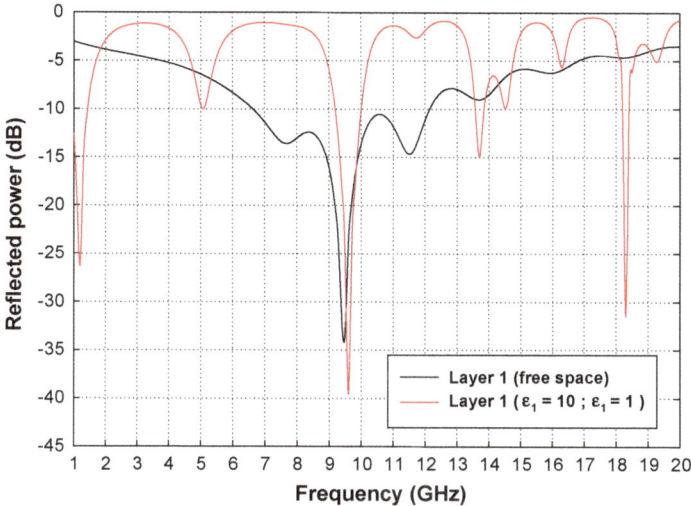

Fig. 27 Effect of plastic envelope on the reflection characteristics of three-layered plastic–plasma–conductor structure

density determines the plasma frequency and hence the resonant absorption band. The resonant frequency, reflection loss (absorption), and its bandwidth depend upon the layer parameters.

Plasma thickness: The effect of plasma thickness on absorption is found to be same as in two-layered plasma–conductor case. Figure 28 shows the effect of plasma thickness on the reflection behavior of four-layered plasma–RAM structure $(N_e = 5 \times 10^{17}$ m^{-3}, $\omega_p = 6.35$ GHz, $V_{en} = 20$ GHz, $d_1 = 1$ cm, $d_3 = 1$ cm, lossy dielectric RAM).

When incident frequency is greater to plasma frequency, the absorption occurs for thin plasma ($d_2 = 5$ cm, 10 cm). However at much higher incident frequency, the absorption peak shifts toward thicker plasma thickness ($d_2 = 50$ cm).

The RAM parameters affect optimum plasma thickness for maximum absorption in the structure. At low incident frequency region, the reflection characteristics of the structures are similar. This is mainly due to reflection by plasma layer, before reaching RAM layer. Once the wave enters into the plasma medium, the RAM parameters will start playing role in affecting the reflection of EM wave.

Figure 29 shows the reflection characteristics *w.r.t* plasma thickness for three-layered (plastic–plasma–conductor) and four-layered plasma–RAM structures. It may be seen that the reflected power *w.r.t.* plasma thickness behavior is similar to that of two-layered structure discussed. It is noted that the reflected power of plasma with and without RAM layer saturates to same value as the plasma thickness increases. The thickness at which reflection saturates is less for four-layered structures (plasma with RAM). The RAM layer decreases the absorption peaks in the reflection characteristics. If the plasma thickness is high the

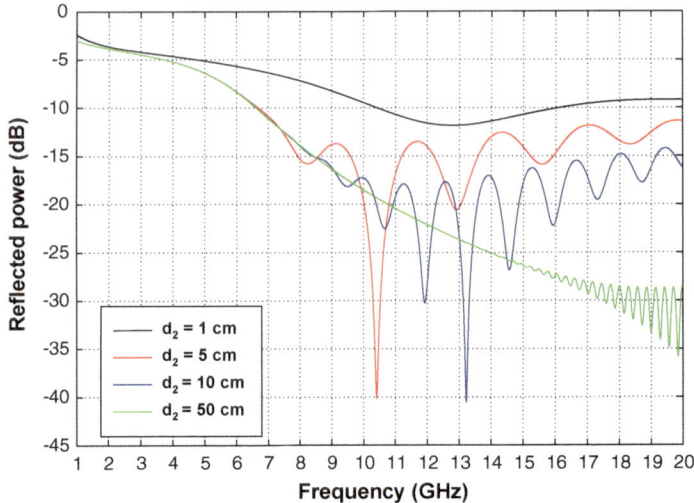

Fig. 28 Effect of plasma thickness on four-layered plasma–RAM structure (RAM-type1)

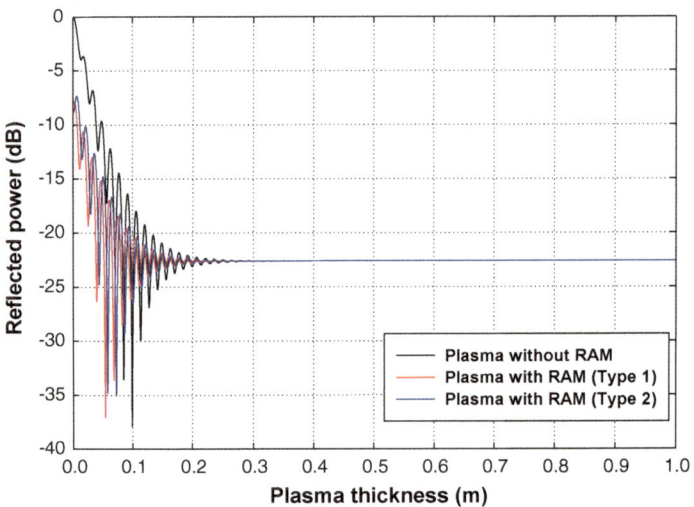

Fig. 29 Comparison of reflected power of three-layered and four-layered plasma structures ($f = 12$ GHz, $N_e = 5 \times 10^{17}$ m^{-3}, $V_{en} = 30$ GHz, $d_1 = 1$ cm, $d_3 = 1$ cm)

role of RAM layer parameters on the reflection characteristics becomes negligible. Thus, one infers that as the plasma thickness increases

- EM wave absorption band shifts toward high incident frequency region
- Optimization of plasma parameters is required to achieve maximum EM wave absorption

- Ripples in reflection characteristics increases
- For thick plasma, the effect of cavity resonance is less
- For large plasma thickness, EM wave absorption is more at high incident frequencies ($\omega \gg \omega_p$)

The increase in $\omega - \omega_p$ (difference between the incident frequency and plasma frequency) results in

- Absorption peak increases and it shifts toward higher plasma thickness
- Reflection loss saturates at higher plasma thickness

The increase in plasma collision frequency results in

- Irregular variation of maximum absorption
- Reflection loss saturates at higher value

Effect of collision frequency: The effect of collision frequency on the reflection characteristics of the four-layered plasma-RAM and two-layered plasma structure are similar. Higher collision frequency shows better absorption. Figure 30 shows the effect of collision frequency on the reflection characteristics of two-layered conductor-backed plasma and four-layered plasma–RAM structure ($N_e = 5 \times 10^{17}$ m^{-3}, $d_1 = 1$ cm, $d_2 = 5$ cm, $d_3 = 1$ cm). If the collision frequency is less, the effect of RAM layer is visible at high incident frequencies ($\omega > \omega_p$). For high collision frequency, the effect of RAM layer parameters is visible in the reflection characteristics (Fig. 30).

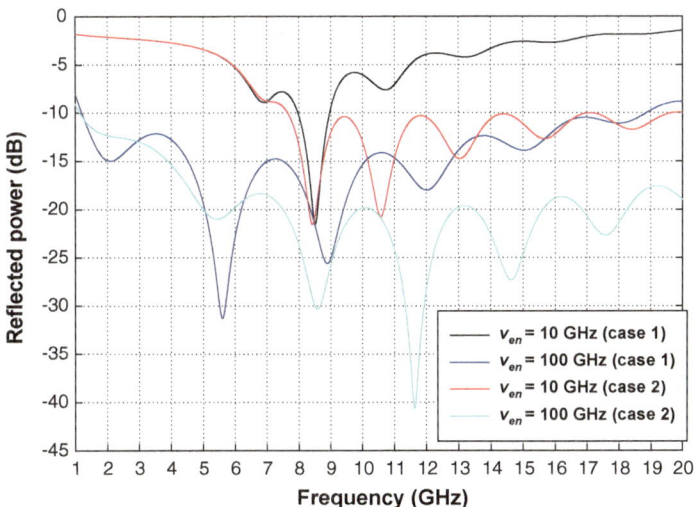

Fig. 30 Effect of collision frequency on reflected power Case 1: plasma without RAM, Case 2: plasma with RAM (Type 1)

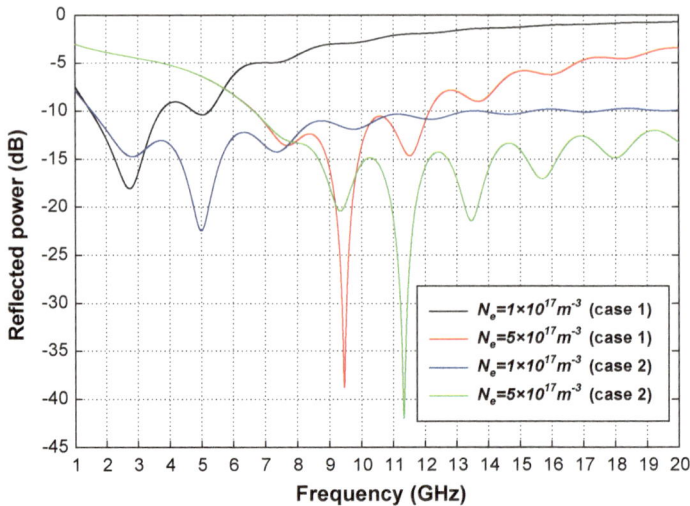

Fig. 31 Effect of electron density on the reflected power Case 1: plasma without RAM, Case 2: plasma with RAM (type 1; $d_3 = 1$ cm)

Effect of electron density (N_e): The plasma electron density plays the same role in both the four-layered and two-layered plasma structures. Figure 31 shows the effect of electron density with $V_{en} = 20$ GHz, $d_2 = 6$ cm, on the performance of two structures. The plasma frequency, obtained from the plasma electron density, determines the resonant absorption band. The RAM layer slightly shifts the resonant absorption band, depending on its parameters.

4.2.4 Effect of Inhomogeneity in Plasma Layer

Here the plasma layer is assumed to be inhomogeneous having specific density profile. The four-layered plasma–RAM structure is considered to analyze the effect of inhomogeneity in plasma layer. Plasma layer is divided into twelve sublayers. The reflection characteristic of the inhomogeneous plasma with linear electron density profile is shown in Fig. 32. The parameters of layer 1 (plastic) and layer 3 (RAM) are given in Table 7. It is to be noted that N_e is the electron density for homogeneous plasma, whereas it is the maximum electron density for inhomogeneous plasma.

It is apparent from Fig. 33 that the plasma inhomogeneity alters the reflection characteristics of the structure. There is only one absorption peak in case of inhomogeneous plasma. Moreover, the inhomogeneity in the plasma shifts the resonant absorption band toward low incident frequency region as compared to homogeneous plasma.

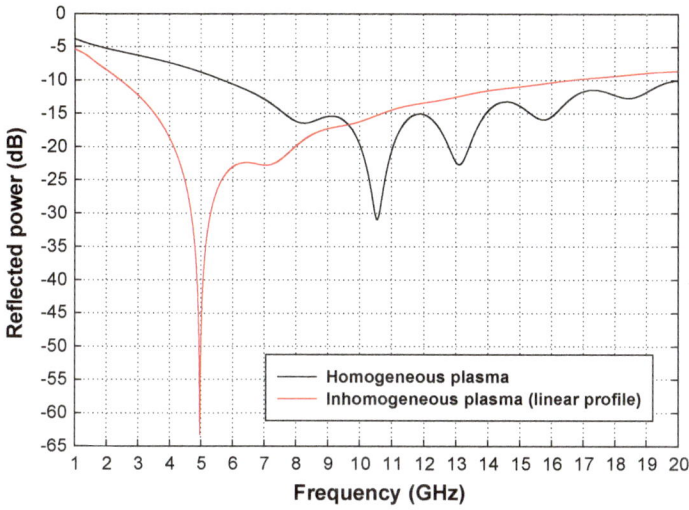

Fig. 32 Effect of inhomogeneity on the reflected power of four-layered plasma-RAM structure; $N_e = 5 \times 10^{17}$ m^{-3}, $V_{en} = 30$ GHz, $d_2 = 5.2$ cm

Table 7 Layer parameters (Fig. 33)

Structure	Thickness (cm)	Parameters
Layer 1 (envelope)	1	$\varepsilon_r = 1$, $\mu_r = 1$ (free space parameters)
Layer 3 (RAM)	1	Type 2 material

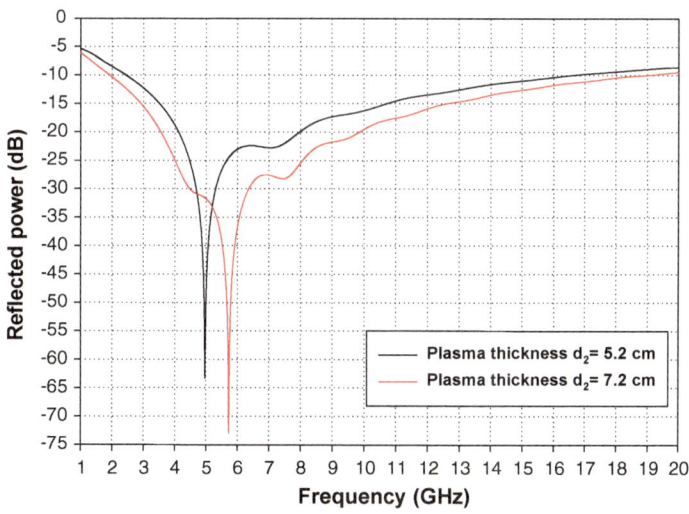

Fig. 33 Effect of plasma thickness on reflected power of four-layered structure having inhomogeneous plasma with linear electron density profile $N_e = 5 \times 10^{17}$ m^{-3}, $V_{en} = 30$ GHz

For homogeneous plasma the maximum EM wave absorption occurs at $\omega > \omega_p$. However, this is not true in case of inhomogeneous plasma. This may be due to the fact that in inhomogeneous plasma, the maximum electron density is near the plasma–RAM interface instead of plastic–plasma interface. In other words, at the plastic–plasma interface, the electron density is small and hence wave with low incident frequency can enter into the plasma and get absorbed.

The linear density profile of the plasma eliminates the abrupt impedance transition at the plasma–air interface. The absorption bandwidth increases due to the inhomogeneity in the plasma. The performance of the structure with linear plasma electron density profile is shown in Table 8.

Figure 33 presents the effect of plasma thickness on the reflected power of four-layered structure having inhomogeneous plasma with linear electron density profile ($N_e = 5 \times 10^{17}$ m^{-3}, $V_{en} = 30$ GHz). It may be observed that for $d_2 = 7.2$ cm the absorption is more and at higher incident frequency with greater bandwidth.

This effect of plasma thickness has been also observed in homogeneous plasma case. However, the bandwidth of absorption decreases at higher plasma thickness.

Figure 34 shows the effect of maximum electron density on the reflected power of four-layered structure having inhomogeneous plasma with linear electron density

Table 8 Performance of inhomogeneous plasma with linear density profile

Nature of plasma medium	Reflected power (dB)	Resonant frequency (GHz)	BW (GHz)
Homogeneous plasma	30.93	10.56	1
Inhomogeneous plasma (linear profile)	63.27	4.97	3.85

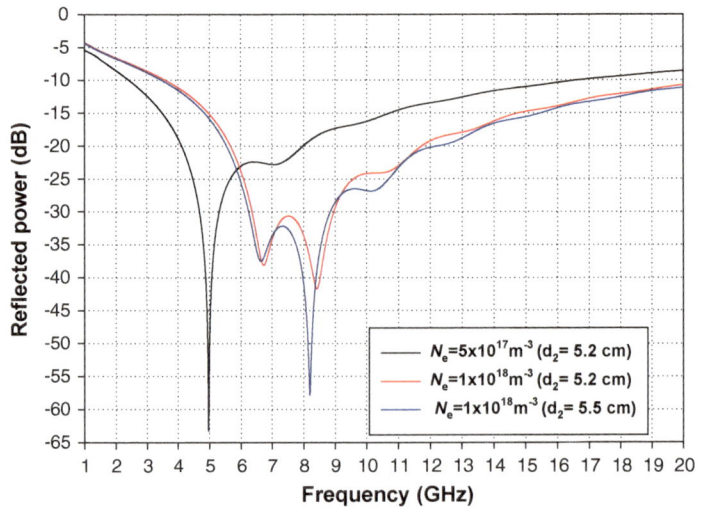

Fig. 34 Effect of electron density on the reflected power of four-layered structure having inhomogeneous plasma with linear electron density profile; $V_{en} = 30$ GHz

profile (V_{en} = 30 GHz). The absorption peak shifts to high incident frequency with greater bandwidth for high electron density. This trend is similar to homogenous plasma case.

In Fig. 35, the effect of collision frequency on the reflected power of four-layered structure having inhomogeneous plasma with linear electron density profile is shown. As the collision frequency increases, the absorption peak shifts slightly toward the lower incident frequency region. Further with optimized plasma thickness the absorption can be increased. This trend is same as in homogeneous plasma case.

One can infer that the inhomogeneity of the plasma is more effective when the plasma thickness is more. The effects discussed above are summarized in Table 9.

Figure 36 compares the reflection characteristics of four-layered plasma–RAM structure with homogeneous and inhomogeneous plasma. Three electron density profiles, viz. linear, parabolic, and exponential are considered. It can be noted that due to electron density variation in the plasma region, the resonant frequency shift toward low incident frequency region. The shift in resonant frequency toward low incident frequency region is less in parabolic and exponential density profile case compared to linear density profile. This may be due to the fact that for linear density profile the electron density at the plastic–plasma interface is less as compared to parabolic and exponential density profile. The less absorption in case of linear density profile is due to the plasma thickness used. The performance of inhomo-geneous plasma with linear density profile can be improved by optimizing the plasma thickness. The effect of various density profiles on the reflection charac-teristics of the structure is tabulated in Table 10.

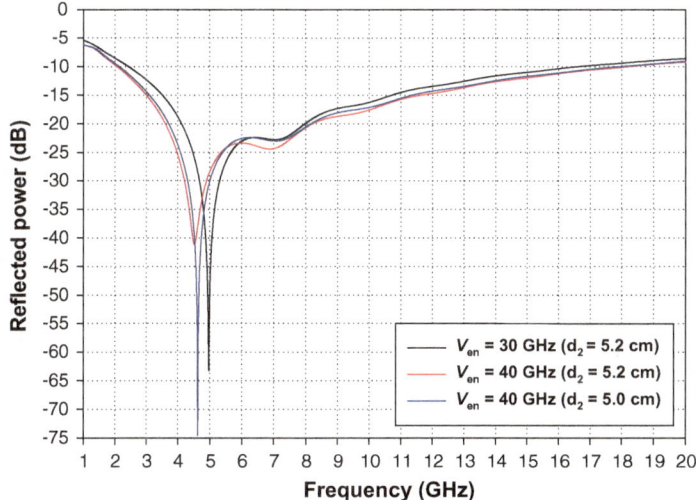

Fig. 35 Effect of collision frequency on the reflected power of four-layered structure having inhomogeneous plasma with linear electron density profile; $N_e = 5 \times 10^{17}$ m^{-3}

Table 9 Performance of the plasma parameters (Inhomogeneous plasma with linear electron density profile)

Plasma parameter	Effects
Thickness (d_2)	Increase in thickness Shifts the absorption band to high frequency region Reflection loss increases Bandwidth increases
Maximum electron density (N_e)	Increase in maximum electron density Shifts the absorption band to high frequency region Bandwidth increases
Collision frequency (V_{en})	Increase in collision frequency Slight Shift in absorption band toward low frequency region Reflection loss increases Slight increase in bandwidth

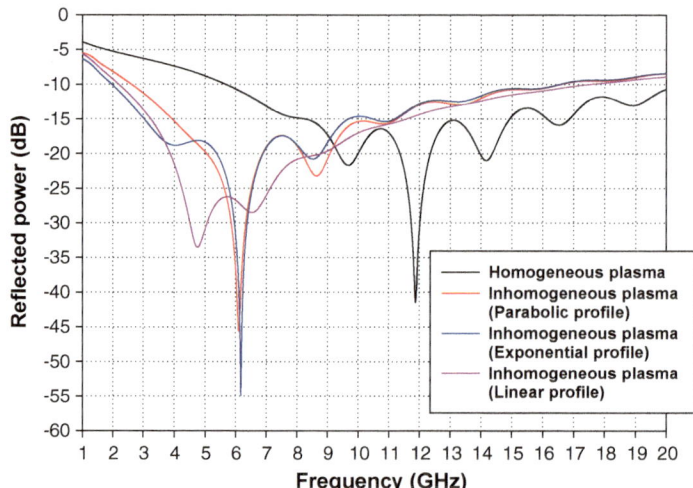

Fig. 36 Comparison of absorption in homogeneous and inhomogeneous plasma structures; $N_e = 5 \times 10^{17}$ m^{-3}, $V_{en} = 30$ GHz, $d_2 = 5.9$ cm

Table 10 Performance of the inhomogeneous plasma

Nature of plasma medium	Reflected power (dB)	Resonant frequency (GHz)	BW (GHz)
Homogeneous plasma	41.47	11.88	0.98
Inhomogeneous plasma (linear profile)	33.49	4.75	4.93
Inhomogeneous plasma (parabolic profile)	45.59	6.1	1.81
Inhomogeneous plasma (exponential profile)	54.55	6.17	1.5

5 Conclusion

Here the impedance transformation method is employed to analyze the reflection characteristics of the bounded plasma. The structure is approximated as a multi-layered dielectric medium. The reflection coefficient is derived for both homogenous and inhomogeneous plasma. The computed results for two-layered conductor-backed plasma structure are validated against the results available in open domain. The reflection performance of the bounded plasma can be improved by including RAM layer and inhomogeneous plasma. The bounded plasma gives rise to cavity resonance effect within the structure. The dependence of reflection characteristics on various plasma parameters are analyzed for plastic–plasma–conductor, and plastic–plasma–RAM–conductor. The conducting layer is assumed to be perfect electric conductor.

The observations made in the study carried out can be summarized as follows: (i) for thick plasma, $\omega_p \ll \omega$ and high collision frequency, greater absorption can be achieved, (ii) for thin plasma, $\omega_p < \omega$ shows more absorption, (iii) EM wave absorption depends on plasma collision frequency, plasma thickness, and RAM parameters, (iv) location of the absorption band mainly depends on plasma frequency, (v) absorption bandwidth essentially depends on plasma collision frequency, (vi) absorption peaks in reflection characteristics is mainly due to cavity resonance effect and electron collisions in the plasma, (vii) inhomogeneity in the plasma layer improves the reflection characteristics, and widens the absorption band, (viii) inhomogeneity in plasma is more effective at large plasma thickness, and (ix) optimization of plasma parameters is required to get improved EM wave absorption.

In this book normal EM wave incidence is considered. The absorption behavior will change for an obliquely incident EM wave. Further the temperature of plasma can change the reflection characteristics of the structure. If this plasma temperature is beyond certain level, the cold plasma model approximation will not be valid. In inhomogeneous plasma, due to different dielectric constant, the incident angle of the propagating EM wave within the plasma layer might vary. This effect is ignored in the present work. However, this method can predict overall picture of reflection characteristics of plasma–RAM structure.

References

Bittencourt, J.A.: Fundamentals of Plasma Physics, 3rd edn. Springer Science and Business Media, New York, 678 p (2004). ISBN: 0387209751

Chaudhury, B., Chaturvedi, S.: Three dimensional computation of reduction in radar cross section using plasma shielding. IEEE Trans. Plasma Sci. **33**(6), 2027–2034 (2005)

Fridman, A., Kennedy, L.A.: Plasma Physics and Engineering. CRC press, New York, 882 p (2004). ISBN: 1560328487

Gruel, C.S., Oncu, E.: Interaction of electromagnetic wave and plasma slab with partially linear and sinusoidal electron density profile. Prog. Electromagn. Res. Lett. **12**, 171–181 (2009)

Hayt, W.H., Buck, J.A.: Engineering Electromagnetics. McGraw-Hill Publishing, New York, 608 p (2011). ISBN: 0073380660

Jenn, D.C.: Radar and Laser Cross Section Engineering, 2nd edn. AIAA Education series, Washington DC, 505 p (2005). ISBN-13: 9781563477027

Mo, J., Yuan, N.: Analytical solution of reflection coefficient microwaves oblique incidence on a nonuniform magnetized plasma slab. In Proceedings of International Conference on Microwave and Millimeter Wave Technology, Nanjing, vol. 4, pp. 1930–1933, April 2008

Shul, R.J., Pearton, S.J.: Handbook of Advanced Plasma Processing Techniques. Springer Science and Business media, New York, 653 p (2000). ISBN: 3540667725

Tang, D.L., Sun, A.P., Qiu, X.M., Chu, P.K.: Interaction of electromagnetic waves with a magnetized nonuniform plasma slab. IEEE Trans. Plasma Sci. **31**(3), 405–410 (2003)

Yuan, C.X., Zhou, Z.X., Sun, H.G.: Reflection properties of electromagnetic wave in a bounded plasma slab. IEEE Trans. Plasma Sci. **38**(12), 3348–3355 (2010)

Yuan, C.X., Zhou, Z.X., Zhang, J.W., Xiang, X.L., Feng, Y., Sun, H.G.: Properties of propagation of electromagnetic wave in a multilayer radar absorbing structure with plasma and radar absorbing material. IEEE Trans. Plasma Sci. **39**(9), 1768–1775 (2011)

Zhengli, H., Ding, J., Chen, P., Zhang, Z., Guo, C.: FDTD analysis of three dimensional target covered with inhomogeneous unmagnetized plasma. In: International Conference on Microwave and Millimeter Wave Technology, Chengdu, pp. 125–128, 8–11 May 2010

Index

© The Author(s) 2017
H. Singh et al., *EM Wave Propagation Analysis in Plasma Covered Radar Absorbing Material*, SpringerBriefs in Computational Electromagnetics, DOI 10.1007/978-981-10-2269-2